g Resources Centre
Library

Electrical Product Safety

About the authors

Jimmy Tzimenakis, BSc., C.Eng., MIEE, has more than 16 years' experience in the testing and approval of electrical products. He is presently the Senior Product Approvals Manager of a multinational manufacturer of consumer electronic products. He has set up an EMC and Safety Test laboratory and introduced (affixed) the CE Marking to hundreds of products.

He is an Approvals Liaison Engineer (ALE) within the BEAB scheme, a member of the steering committee of CE Club (Wales), a member of the British Radio and Electronic Equipment Manufacturers' Association's (BREMA's) Safety and EMC sub-committees, a member of BSI's EMC sub-committee GEL 210/7; he is also a member of the IRCA (International Register of Certified Auditors).

Dave Holland, MITSA, MBA, is the Trading Standards Manager at Cardiff County Council. He has responsibility for all aspects of Trading Standards legislation and has spent many years advising manufacturers on the implications of the 'New Approach Directives'. He has over 14 years' experience in Trading Standards and sits on a variety of industry committees. He is also a LACOTS (Local Authority Co-ordination on Trading Standards) adviser on issues arising from European Product Safety Directives and is a member of the Institute of Trading Standards. He has a Masters in Business Administration and is a member of the steering committee of CE Club (Wales).

Both authors are regular speakers at local and national seminars organized by various clubs and organizations.

Electrical Product Safety: A Step-by-Step Guide to LVD Self-Assessment

Jimmy Tzimenakis and Dave Holland

UHI Millennium Institute
10600541

Newnes

OXFORD AUCKLAND BOSTON JOHANNESBURG MELBOURNE NEW DELHI

Newnes
An imprint of Butterworth-Heinemann
Linacre House, Jordan Hill, Oxford OX2 8DP
225 Wildwood Avenue, Woburn, MA 01801–2041
A division of Reed Educational and Professional Publishing Ltd

 A member of the Reed Elsevier plc group

First published 2000

© J. Tzimenakis and D. Holland 2000

All rights reserved. No part of this publication may be
reproduced in any material form (including photocopying
or storing in any medium by electronic means and whether
or not transiently or incidentally to some other use of
this publication) without the written permission of the
copyright holder except in accordance with the provisions
of the Copyright, Designs and Patents Act 1988 or under
the terms of a licence issued by the Copyright Licensing
Agency Ltd, 90 Tottenham Court Rd, London, England W1P 9HE.
Applications for the copyright holder's written permission
to reproduce any part of this publication should be
addressed to the publishers.

British Library Cataloguing in Publication Data
A catalogue record for this book is available from the British Library.

ISBN 0 7506 4604 7

Library of Congress Cataloguing in Publication Data
A catalogue record for this book is available from the Library of Congress.

Typeset by David Gregson Associates, Beccles, Suffolk
Printed and bound in Great Britain by MPG Books Ltd, Bodmin, Cornwall

Contents

List of tables	x
List of figures	xi
Preface	xiii
Acknowledgements	xv
Introduction	xvii

Part 1
CE Marking and Regulations

1	CE Marking	3
	1.1 The CE Marking concept	3
	1.2 The CE Marking Directives	4
	1.3 National implementation	5
	1.4 Summary	5
2	The Low Voltage Directive – introduction and overview	6
	2.1 Background	6
	2.2 Scope of the LVD	7
	2.3 Exemptions	7
	2.4 The core requirements	7
	2.5 Who is affected?	9
3	Market surveillance	10
	3.1 Legal obligations	10
	3.2 Administration of the New Approach Directives	11
	3.3 Policing the LVD	11
	3.4 The safeguard clause	12
	3.5 The enforcement bodies	12
	3.6 Enforcement 'tools'	13
	3.7 Criminal offences	13
	3.8 Defence provisions	14
	3.9 Summary	14

4	'Due diligence'		15
	4.1	How Product Safety law works in the UK	15
	4.2	Reasonable precautions and due diligence – the concept	16
	4.3	Reasonable precautions and due diligence – a good practice guide	17
		4.3.1 Assess the risk	17
		4.3.2 Establish a plan	18
		4.3.3 Write down a solution	18
		4.3.4 Operate the system	19
		4.3.5 Review the system	19
	4.4	Reasonable precautions and due diligence – a good practice example	19
	4.5	Conclusions	21

Part 2
The practical solution

5	Standards – an overview		25
	5.1	Product specific standards	25
	5.2	European standards	26
	5.3	When product specific standards do not exist	29
	5.4	Summary	29
6	Testing for safety		30
	6.1	To test or not to test?	30
	6.2	Do I need a Certification mark?	31
	6.3	How do I get a Certification mark?	31
	6.4	Where to test and how long will it take?	32
	6.5	Which standards to test against?	34
	6.6	How much will it cost?	34
	6.7	How do I ensure compliance?	34
	6.8	Summary	36
7	Self-evaluation – route to compliance		37
	7.1	Step 1 – Finding an applicable standard/regulation	37
	7.2	Step 2 – Understand the fundamental safety requirements	38
	7.3	Step 3 – During the design stages	38
	7.4	Step 4 – Test equipment considerations	39
	7.5	Step 5 – Decision of Safety Critical parts	39
	7.6	Step 6 – Prototype testing	40
	7.7	Step 7 – Creating an Evaluation Report	40
	7.8	Step 8 – Testing the final product	40
	7.9	Step 9 – The final act	41
	7.10	Summary	41
8	Fundamentals of product safety		43
	8.1	Protection against electric shock	45

	8.2	Protection against fire	46
	8.3	Protection against mechanical hazards	46
	8.4	Protection against temperature rises	47
	8.5	Protection against energy hazards	47
	8.6	Protection against dangerous chemicals	48
	8.7	Other hazards	48
	8.8	Summary	48
9	Choosing Safety Critical components	49
	9.1	Component approval/selection – options	50
	9.2	Using approved components	51
	9.3	Approval marks	52
	9.4	Safety Critical components – examples	53
	9.5	Summary	55
10	Designing for safety	56
	10.1	Definitions	57
	10.2	Equipment classification	57
		10.2.1 Class I	58
		10.2.2 Class II	60
		10.2.3 Class III	60
	10.3	Insulation	61
		10.3.1 Basic insulation	61
		10.3.2 Supplementary insulation	61
		10.3.3 Double insulation	62
		10.3.4 Reinforced insulation	62
		10.3.5 Bridging insulation	63
		10.3.6 Creepage and clearance	64
	10.4	Construction – wiring	67
		10.4.1 Internal wiring	67
		10.4.2 Internal wiring – mechanical fixing	67
		10.4.3 External wires and mains plugs	68
		10.4.4 Internal wiring – insulation	71
		10.4.5 Withdrawal of mains plug	72
	10.5	External construction	72
		10.5.1 Mechanical strength	74
		10.5.2 Mechanical stability	75
	10.6	Resistance to fire	75
	10.7	Electrical connections and mechanical fixings	76
	10.8	Components	77
	10.9	Temperature	78
		10.9.1 Heating under normal operating conditions	78
		10.9.2 Heating under fault conditions	79
	10.10	Dielectric strength	81
	10.11	Insulation resistance	82
	10.12	Earth leakage current	83
	10.13	Measurement and measurement equipment	83

	10.14 Protection of service personnel	84
	10.15 Other hazards	84
	10.15.1 Spillage	84
	10.15.2 Overflow	84
	10.15.3 Liquid leakage	85
	10.15.4 X-rays and cathode-ray tubes for TV and computer monitors	85
	10.15.5 Ultraviolet radiation	85
	10.15.6 Microwave radiation	85
	10.16 Labels	85
	10.17 Markings	87
	10.18 User instructions	88
11	Compiling an Evaluation Report	89
	11.1 Produce specific reports	89
	11.2 Own design reports – contents	90
	11.3 Product Evaluation Report – example	92
	11.4 Summary	112

Part 3
Preparing the documentation

12	Technical File	115
13	Declaration of Conformity	118
	13.1 Declaration of Conformity – contents	118

Part 4
Setting up production control

14	Production control	123
	14.1 Essential 'tools'	123
	14.1.1 Safety Critical component lists	124
	14.1.2 Safety Critical assemblies	124
	14.1.3 Preparing operation standards	126
	14.1.4 Preparing a test manual	126
	14.2 Production control procedures	126
	14.3 Engineering change control	127
	14.4 Control of purchasing	128
	14.5 Confirmation at incoming goods area	128
	14.6 Adherence to operation standards	128
	14.7 Routine testing	129
	14.8 Testing at the end of line (final inspection)	129
	14.8.1 Earth continuity	130
	14.8.2 Dielectric strength (flash) test	130
	14.8.3 Insulation resistance	131
	14.8.4 Earth leakage current	131

	14.9 Records	132
	14.10 Audits	132
	14.11 Summary	133
15	Factory control guidelines	134
	15.1 Incoming goods area	134
	15.2 Stores	135
	15.3 Assembly lines	135
	15.4 End-of-line tests (final safety)	136
16	Quality processes and the LVD	138
	16.1 Safety audits	138
	16.2 Quality system	139
	16.3 Procurement of Safety Critical materials	139
	16.4 Manufacturing control	141
	16.5 Quality assurance functions	142
	16.6 Due diligence procedure	142
	16.7 Rework/Market recall	143
	16.8 Summary	144

Appendices

1	List of European Union Directives	146
2	List of Harmonized safety standards	148
3	UK Notified Bodies under the Low Voltage Directive	156
4	Other European Certification Bodies/Testing Laboratories	158
5	List of test equipment for electrical safety testing	161
6	Example of the contents of a Harmonized safety standard	163
7	Insulation requirements between parts – guidance examples	175
8	Insulation types, electrical connections and examples	179
9	Creepage distances and clearances – measurement guide	181
10	Test circuit for measuring 'touch currents'	186
11	Test instruments	187
12	Graphic symbols	191
13	Examples of an EC Declaration of Conformity	193
14	Procedure for handling Safety Critical components and operations	195
15	Engineering Change Note – example and completed document	196
16	Production Change Request – example	199
17	End-of-line (final safety) tests	201

18	Useful addresses	202
19	Glossary of terms	204
20	Templates	206
Index		229

List of tables

Table 3.1	Enforcement options	13
Table 8.1	Summary of basic safety principles	44
Table 8.2	Summary of some special safety hazards	45
Table 9.1	Examples of Safety Critical components	53
Table 10.1	Example – values of insulation thickness, recommended distances and test levels	63
Table 10.2	Definitions of pollution degree	66
Table 10.3	Operational, basic and supplementary insulation	66
Table 10.4	Some typical maximum permissible temperature rises	80
Table 10.5	Dielectric strength test voltages	82
Table 10.6	Typical insulation resistance values	82
Table 14.1	Example of a Safety Critical components list	124
Table 14.2	Example of a Safety Critical assembly list	125
Table 14.3	Example of an Operation Standard	125
Table 14.4	Dielectric strength test – values based on consumer product standards	131
Table 14.5	Example of internal audit checklist	133

List of figures

Figure 4.1	'Due diligence' activities – example flowchart	20
Figure 4.2	Operation procedure – example	21
Figure 4.3	Safety Critical operations index – example	22
Figure 5.1	Route to compliance with the LVD	28
Figure 6.1	Options for testing	33
Figure 10.1	Example of a good earthing connection	59
Figure 10.2	Example of a double insulated wire	62
Figure 10.3	Example of a safety component bridging the HV/LV barrier	64
Figure 10.4	Print side HV/LV circuitry separation	65
Figure 10.5	Example – internal wiring dressing	68
Figure 10.6	Example of an approved mains lead	69
Figure 10.7	Example of an approved mains plug	69
Figure 10.8	Examples of enclosure openings	74
Figure 10.9	Example of fuse marking on PWB	77

Preface

This is a practical and easy-to-follow guide aimed at helping manufacturers of electrical products to understand the relevant basic safety principles, to perform self-assessment of their products, to create their own customized safety reports and to meet the requirements of the Low Voltage Directive. However, solely following the guidelines as given in this book may not be sufficient to provide due diligence defence. The requirements of the Directive must be followed, and the manufacturer will need to decide appropriate procedures and confirmation methods depending on production quantities, product range and ongoing innovation. The authors of this guide have different backgrounds, yet by combining their practical and technical experiences with the legislative and enforcement requirements, they have attempted to give the reader balanced views and opinions which will be suitable for most manufacturers.

Being members of their regional 'EU Directive Awareness Club' since 1994, the authors have attended many club seminars and taken part in many discussions on how to meet the requirements of a variety of EC Directives. In those seminars, speakers with technical, legislative or administrative expertise would explain in detail what need be done, and proceed to offer suggestions on the best approach to meet the requirements of various pieces of European legislation.

The response of the club members to these presentations often varied. Some would be very interested; these tended to be representatives of large-scale manufacturers of electrical products with the personnel and financial resources to enable speedy compliance. They would indicate what actions they had already taken to meet the requirements and happily talk about the vast sums of money invested. Others would display clear signs of frustration and sometimes even anger. 'There is no way I can comply with that' was a commonplace reaction. Such people tended to be the representatives of small companies or sole traders who employed few people and produced either 'one-offs' or a small range of products every month or year. They would shake their heads, clearly with the unspoken (sometimes even voiced) thought, 'This will finish us.' Others were apathetic and obviously intended to carry on as normal and hope 'it would go away'.

However, one clear theme did emerge from such small-scale manufacturers. Many would argue with the experts that because they produced only a few units each year, this legislation should not include them. They complained that the examination fees levied by the test houses made it impossible to test and remain profitable. Many looked for a way to reduce testing requirements and paperwork without compromising the objective which the Directives set out to achieve. For, in spite of the complaints, most of the manufacturers appreciated that EC Directives were designed and prepared for a good reason.

The small-scale manufacturer is finding it hard to come to terms with meeting all of the requirements of the 'New Approach Directives'. Quite simply they need help – help that is easy to understand, financially sound and above all practical. This guide sets out to do just that. It aims to present the requirements of the Low Voltage Directive in a format that is simple to understand, and offers practical advice on how the essential requirements may be met without incurring unnecessary expense and bureaucracy.

Although every effort has been made in this book to present the reader with a simplified process for product evaluation and self-assessment, the authors feel it is necessary to point out that the information provided in Chapter 7 (Self-evaluation – route to compliance), Chapter 10 (Designing for safety) and Chapter 11 (Compiling an Evaluation Report) is very general, and on its own will not be sufficient to warrant complete and accurate product evaluation as required by the Low Voltage Directive. These chapters have been compiled by taking the core requirements from a number of safety standards which is then presented as a means of raising the reader's awareness of their existence. Only Harmonized Product Specific Standards will provide sufficient details of the testing methods and pass/fail criteria necessary for product acceptance.

Jimmy Tzimenakis
Dave Holland

Acknowledgements

Extracts from British Standards are reproduced with the permission of BSI. Complete editions of the standards can be obtained by post from BSI Customer Services, 389 Chiswick High Road, London W4 4AL.

The authors also gratefully acknowledge the kind contributions made by Ken Allen, Norman Lloyd, Simon Barrowcliff and Tony Leathart for the advice on technical issues, presentation and grammar.

Reference

Burt, P.H.G. (1995) Designing for equipment safety – a practical guide for the European Market, *ERA Report 95-0035*, Leatherhead, UK: ERA Technology Ltd, July, 176 pp.

Our special thanks to Janice and Hilary, our wives, for their support during the preparation of this guide.

Important disclaimer

This publication is intended to provide accurate information on the subject matter covered and guidance in relation to product testing methodology. However, it does not obviate the need to seek appropriate professional, technical or other competent advice in particular circumstances. Readers testing products or entering into transactions on the basis of the contents should therefore not rely exclusively on testing procedures contained herein. Accordingly no warranty is made or undertaking is given that products evaluated using the methods contained herein will or will not comply with any legislative requirements.

Copyright in all statutory and other materials resides in BSI or other relevant body. Copyright material is vested in the authors.

Introduction

The guide is divided into four parts:

- The first part examines the Regulations, their enforcement in the United Kingdom (UK) and the concept of due diligence.
- The second and most detailed part will take the reader through the process of product self-evaluation and report compilation.
- Part 3 deals with the documentation, i.e. how to compile a Technical File and how to prepare a Declaration of Conformity.
- Finally, Part 4 explains how to set up factory and production control systems.

In the guide, the reader will find:

- the concept of CE Marking
- an overview of the Low Voltage Directive and the implementing UK regulations
- an explanation of how the law is enforced and the sanctions for non-compliance
- the concept of due diligence
- explanations on safety standards
- the route to self-evaluation
- the fundamentals of product safety
- advice on choosing the correct Safety Critical parts
- advice on designing for safety and product assessment
- advice on creating a customized Product Evaluation Report
- guides for compiling a Declaration of Conformity and a Technical File
- a practical approach for establishing and putting in place a Production Control system for safe manufacturing
- integration of safety and quality.

Part 1

CE Marking and Regulations

Chapter 1

CE Marking

1.1 The CE Marking concept

What is CE Marking? Why is it needed? Who needs to use it? How do they obtain it?

Answering these questions continues to present difficulties for manufacturers of electrical and electronic goods, yet understanding the concept of CE Marking is fundamental to successful marketing in Europe. In this chapter we try to answer the questions listed above and illustrate the importance of CE Marking.

It should be noted at the outset that CE Marking is primarily concerned with promoting free trade and not the creation of additional legislative controls. The European Union (EU) has been working towards the concept of free trade for many years. It was one of the major influences behind the signing of the Treaty of Rome in 1957. The Treaty committed the member states of the original European Economic Community (EEC) to work toward this goal.

This was a radical concept and required each country to discard its own laws, in specific areas, and replace them with a harmonized set of European rules. This process has been a slow one, taking over 40 years so far, and there is still a long way to go. However, considerable progress has been made in harmonizing the laws pertaining to product safety. The decision to harmonize product safety laws was agreed under Article 100A of the Treaty of Rome and thus the resulting Directives are known as the 100A or New Approach Directives.

The aim of the product safety Directives is simple. If a product can satisfy the provisions of an appropriate harmonized law, then that product must be allowed free passage throughout the EU without the need for further certification or testing.

Manufacturers are able to demonstrate and declare compliance with these conditions by using CE Marking. By affixing the CE Mark, the manufacturer is making a visible statement that his equipment meets all requirements of the relevant Directives. It is important for manufacturers to understand that they themselves are responsible for ascertaining which Directives are applicable to their products. If they fail to apply a relevant Directive, they may face legal action.

1.2 The CE Marking Directives

A full listing of the CE Marking Directives is given in Appendix 1. Of primary importance to the readers of this guide are:

- the Low Voltage Directive 73/23/EEC as amended by 93/68/EEC
- the EMC Directive 89/336/EEC as amended by 91/263EEC and 93/68/EEC
- the Machinery Directive 89/392/EEC as amended by 91/368/EEC, 93/44/EEC and 93/68/EEC.

These Directives, along with the others listed in Appendix 1, all have certain 'standard' elements. These include:

- the scope of the Directive; this contains details of the range of products subject to the controls listed in the Directive, and perhaps more importantly those products which are exempt from its requirements
- a requirement, placed upon member states of the European Community, to ensure that only safe products are allowed into the European market place. This requires member states to consider internal control measures and policing mechanisms
- a safeguard clause to ensure that any unsafe products are prevented from entering the market place or, if any unsafe products are detected, they must be immediately withdrawn from the market
- a listing of certain essential safety requirements which form the core safety elements of a Directive. In the case of the Low Voltage Directive (LVD), these are called the 'safety objectives'
- a statement linking harmonized performance standards to the essential safety requirements and establishing the concept of the 'presumption of conformity'. If an electrical product meets the requirements of a Harmonized standard which covers all the relevant safety related issues contained in the LVD, it is presumed to conform to the 'safety objectives'
- conformity assessment requirements. These differ across the Directives and it is important to read each one carefully. Some of the Directives allow a manufacturer to self-declare compliance, while others require third party involvement. The LVD is one of the former. Manufacturers may use the standards route outlined in Figure 6.1 or manufacture their product in conformity with the essential requirements
- CE Marking requirements. Once again these differ across the Directives with regard to the CE Mark and other additional marks.

1.3 National implementation

The CE Marking Directives harmonize the laws of member states in order to remove barriers to trade. A Directive is simply an instruction, issued by the Council of the European Community, telling a member state to transpose the contents of the Directive into national law; failure to do so would constitute a breach of community law.

It is important to understand that the Directives themselves have no effect on individual manufacturers. The manufacturers' duty is to comply with the transposed national requirements, which, in theory, replicate the provisions of the Directive, with some additional domestic elements such as enforcement and sanctions for non-compliance. At present, there are some 25 pieces of legislation in the UK that transpose and implement the requirements of the European product safety Directives.

1.4 Summary

The CE Marking Directives have given the EU a high-profile method of indicating a product's compliance with safety and other requirements. The European market place is slowly becoming accustomed to the new requirements and is developing an understanding of CE Marking. As a result, businesses are frequently including CE Marking requirements in their purchasing contracts. At the same time, enforcement of the law is becoming more active and prosecutions of non-compliant suppliers and the banning from sale of unsafe products are becoming commonplace.

Today, in order to continue doing business within the EU, manufacturers are being forced by legislation, market surveillance mechanisms and consumer demand to undertake some type of compliance testing and/or verification of their products. CE Marking is no longer an option, it has become a vital component of a producer's business.

Chapter 2

The Low Voltage Directive – introduction and overview

2.1 Background

One of the first Directives adopted by the European Union (EU) to harmonize the laws of member states was European Commission Directive 73/23/EEC, the Low Voltage Directive (LVD). In its original format, the LVD did not provide for the use of CE Marking. However, as the concept of the Single Market developed and other Directives were created, it became apparent that the LVD needed to be revised.

On 22 July 1993, Directive 93/68/EEC ('the CE Marking Directive') was adopted. This Directive harmonized the rules relating to the use of CE Marking for all the product safety Directives. It also amended the Low Voltage Directive in respect of the procedures for conformity assessment. These amendments have been implemented into national laws across the EU, e.g. the legislation now applicable in the UK is the Electrical Equipment (Safety) Regulations 1994 (ESSR).

The LVD is based on the following concept:

- Electrical equipment can only be placed on the market if it does not present a hazard to people, domestic animals and property.
- To meet this aim, only electrical equipment which satisfies all the requirements of the modified LVD is entitled to free circulation through the European Community.

The responsibility for achieving this objective lies with each member state, and is considered further in Chapter 3.

2.2 Scope of the LVD

The LVD applies to all electrical equipment (with some minor exceptions) which is designed or adapted for use between 50 V and 1000 V AC or between 75 V and 1500 V DC. These voltage ratings refer to the voltage of the electrical input or output, not to any internal operating voltages. The Low Voltage Directive is applicable to equipment intended to be used in both domestic and workplace environments, and it is estimated that 80% of all electrical equipment used falls within the scope of this Directive.

Note: *Equipment operating outside the voltage limits given above, may still be required to comply with other legislation.*

While the LVD is primarily concerned with electrical equipment, some '*electrical components*' may also be classed as electrical equipment. Products intended to be incorporated into other electrical equipment, such as transformers and motors, are examples of components covered by the LVD. These component parts must satisfy the requirements of the LVD and bear the CE Marking.

Another group of products covered by the LVD is cable management systems, such as plastic conduits, trunking, etc. While some national governments do not believe this to be a correct interpretation of the intention of the Directive, the European Commission has reconfirmed in their latest guidance that these products must carry CE Marking.

2.3 Exemptions

It is important to note the products which are excluded from the LVD, namely:

- those that are covered by other community directives, such as the use of electrical equipment in explosive atmospheres
- plugs and sockets for domestic use
- electric fence controllers
- specialized electrical equipment intended to be used on ships, aircraft or railways. These must comply with other safety provisions
- electrical equipment for supply outside the EU.

2.4 The core requirements

All electrical equipment subject to the Low Voltage Directive *must*:

- be safe. The Directive covers all risks arising from the use of the product, be they electrical, physical or mechanical

8 Electrical Product Safety

- be constructed in accordance with principles generally accepted within *member states* as constituting good engineering practice in relation to safety matters

Note: *A product is presumed to be safe and therefore compliant if it is manufactured in accordance with a published performance standard. To this end, the Directive recognizes a heirarchy of standards for demonstrating compliance. Initially, the Directive suggests the use of a harmonized European standard, such as EN 60950, which covers the safety of Information Technology Equipment (ITE), EN 60335 which covers the safety of household appliances and EN 60065 the standard for broadcast equipment. If the product meets the requirements of a Harmonized standard, it is presumed to be compliant in terms of safety.*

If a Harmonized standard does not exist for the product, an international standard may be used, such as those drawn up by the International Electrotechnical Commission (IEC). Where neither of these standards exists, a national standard may be used.

The standards referred to above provide a presumption of conformity for any equipment manufactured in compliance with their requirements. However, manufacturers are not compelled to use these standards and can build their products against the safety objectives listed in the Directive. If manufacturers chose to do so, they must demonstrate clearly in their technical documentation, their method of ensuring compliance. Clearly it is advantageous and less time consuming to use published performance standards, but it is not mandatory.

- bear the CE Marking which must be at least 5 mm high and be affixed legibly and indelibly to one (or any combination) of the following:
  - the apparatus itself or where this is not feasible or technically possible
  - the packaging for the apparatus
  - the instructions for use
  - the guarantee certificate

Note: *Marks other than CE Marking (for example, an approval mark from a Certification Body) may appear on or with the equipment but they cannot be used to declare compliance with the Low Voltage Directive.*

- have an EC Declaration of Conformity issued by the manufacturer or his authorized representative within the Community, the content of which is considered in Chapter 13
- have Technical Documentation relating to the product prepared and available for external inspection. It must include sufficient detail to demonstrate that the product complies with the Directive. This documentation is often used by the enforcement authorities to assess the conformity of the electrical equipment against the requirements of the Low Voltage Directive, the content of which is considered in Chapter 12

- be manufactured in accordance with a programe of Quality Assured production.

Note: *The information given above is only an overview of the Low Voltage Directive. Details of the exact requirements can be found in the text of the EC Directive.*

2.5 Who is affected?

It is in this consideration that the Directive and some of the national laws implemented in member states can differ. The Directive uses the term 'placing on the market'. A product is placed on the Community market when it is made available for the first time for sale or distribution. Thus the only person who can place a product on the European market is either the manufacturer, his authorized representative or the Community or the person who first distributes the product in the Community, or an importer.

However, due to linguistic subtleties, the transposed versions of the Directive differ in some of the member states. In the UK for instance, all persons who supply electrical equipment during their course of business are covered by the UK regulations. Therefore the UK regulations affect not only:

- manufacturers
- authorized representatives
- importers into the European Union.

But in addition they cover:

- wholesalers and distributors
- retailers and hirers.

The responsibility for ensuring that electrical equipment is safe will depend upon an individual's position in the supply chain. The manufacturer will obviously carry the greatest burden, but other professionals in the supply chain must be aware of the requirements of the LVD.

Chapter 3

Market surveillance

Market surveillance is an important aspect of the new approach concept. It is based on the desire to provide a high level of protection for users of a given product be it a computer, teddy bear or the most complex of machines. The member states of the EU have therefore established bodies to monitor the market place for unsafe products and have authorized these bodies to take appropriate action where necessary.

For the new approach concept to work, the member states must ensure that their surveillance bodies have the appropriate resources and expertise to ensure the single market operates fairly, and that consumer safety is not compromised. Unfortunately there is a distinct lack of uniformity on this topic with some countries taking a more active role than others.

3.1 Legal obligations

All manufacturers have a legal and moral duty to produce safe goods. If this basic requirement is contravened, then manufacturers, their authorized agents, or importers into the EU may find themselves subject to action in the courts. This may take the form of:

- criminal action – such as prosecution and forfeiture of goods and/or
- civil litigation – being sued by parties aggrieved by the marketing of the product.

The criminal liabilities of a manufacturer arise from the requirements of the national laws that implement the LVD, such as (in the UK) the Electrical Equipment (Safety) Regulations (EESR). The legislation is criminal, i.e. backed by personal fines and/or the possible imprisonment of an organization's officers. Surveillance bodies are appointed to administer its demands.

The civil liabilities of a manufacturer arise from the law of contract and also from the requirements of the Product Liability Directive 85/374/ EEC, which is applicable to all products covered by the LVD and other new approach directives. The product liability regime requires manufacturers to make safe products or face substantial penalties and compensation claims from aggrieved parties.

3.2 Administration of the New Approach Directives

In most member states, it is the role of central government to produce legislation that will implement the EC Directives into national law. Most member states allocate the responsibility for the administration of the new approach laws to a particular government department, which is commonly referred to as the 'lead department' or 'competent authority'. The lead department is required to:

- liaise with Europe
- appoint or remove Notified Bodies
- liaise with the enforcing authorities
- interpret the regulations.

In the case of the LVD in the UK, the duties listed above are carried out by the 'Standards and Technical Regulations Directorate' of the Department of Trade and Industry.

3.3 Policing the LVD

Article 2 of the LVD requires all member states to monitor the production and supply of electrical equipment in their geographical areas. If unsafe electrical products are detected (i.e. which might present a risk to the safety of individuals), member states must take action to remove those products from the market. The aim of the single market is the free movement of goods, but this ideal only applies if the goods in question are safe.

Market surveillance has two distinct parts.

The first part is the activity of monitoring the market for unsafe goods. This is normally done through visits to manufacturers, sampling their products and checking the requisite documentation such as Technical Files and the Declaration of Conformity.

The second part involves the enforcement agencies taking appropriate corrective action to ensure conformity when problems are detected. Such action can be taken against the product and the person who placed the product on the market. Enforcement 'tools' are discussed in 3.6 below.

3.4 The safeguard clause

Article 9 of the Directive requires that if any member state takes action to prohibit the supply or free movement of any electrical equipment, it must inform the European Commission and the other member states of its decision and the reasons behind it.

3.5 The enforcement bodies

The LVD does not stipulate how member states should administer its provisions; consequently there is some inconsistency in enforcement across the EU. In an attempt to rectify this issue, the enforcement agencies in the member states now subscribe to a forum called PROSAFE, the Product Safety Enforcement Forum of Europe. The forum was created to develop operational understanding and trust between enforcement officials. It aims to allow enforcement officers to develop policies on market control, inspection, sampling and co-operation. PROSAFE is in its infancy, but it has the potential to facilitate market inspection at a Community level.

To illustrate how the LVD is enforced by a member state, the remainder of this chapter examines the system operated in the United Kingdom.

The LVD is implemented into UK legislation by the Electrical Equipment Safety Regulations (EESR). The legislation places a positive duty on the UK's Trading Standards Service to administer the provisions of the Directive in a given geographical area. At present there are over 200 Trading Standards Departments in the UK. Consistency of enforcement is ensured through liaison with their co-ordinating body LACOTS (Local Authority Co-ordination on Trading Standards)

Trading Standards Officers have a wide range of powers available to them to enforce the CE Marking Directives and these are examined in 3.6 below.

One of the major aims of the trading standards enforcement strategy is to provide advice to local manufacturers and importers. The trading standards service operates the Home Authority principle, which means that manufacturers only have to deal with their local trading standards department which takes responsibility for the activities of those companies in its geographical area.

3.6 Enforcement 'tools'

Table 3.1 Enforcement options

Tool	Function
Compliance notice	Used on CE Marked products which are not fully compliant with the Directive, e.g. defective documentation.
Suspension notice	Used by an Enforcement Officer to prevent the supply of goods for up to 6 months. The notice 'freezes' the goods in the supply chain. Physical movement of the goods is permitted as long as the Enforcement Officer is aware of it.
Power of seizure	An Enforcement Officer can seize and detain apparatus, records and other information which he believes to be necessary to prove any offences.
Written cautions	This is a 'warning' that the Enforcement Officer can issue when a company has committed an offence serious enough to warrant prosecution, but it is judged not to be in the public interest to prosecute. Written cautions are not used lightly – they are only issued when a conviction would be likely, or if the offender admits the offence and agrees to be cautioned. Cautions can be cited in court in any future proceedings as if they were a previous conviction.
Complaints	This is a tool that gives an Enforcement Officer the option to ask a court to issue a Forfeiture Order for offending items. This allows products to be removed from the market without prosecuting either the owner or the importer of the goods.
Prosecution	If there has been deliberate disregard of the Regulations, or if advice has been ignored, the Enforcement Officer may have no other option than to prosecute.

3.7 Criminal offences

The LVD does not specify any offences nor does it specify any penalties for non-compliance with its requirements. The sanctions to be used against non-compliant products are left for each member state to decide.

In the UK, the EESR describe a number of offences. These include:

- a failure to meet the principal safety objectives
- misuse of the CE Marking
- failure to retain documentation
- issue of incorrect documentation
- giving misleading information to enforcement officers

- obstruction of enforcement officers
- contravention of enforcement notices.

The EESR allow the enforcing authorities to prosecute suppliers if they fail to meet the statutory requirements. The maximum penalties for non-compliance are a £5000 fine and/or 3 months' imprisonment per offence.

3.8 Defence provisions

Guidance on claiming a defence against any charges brought under the EESR may be found in Section 39 of the Consumer Protection Act 1987. To be successful in defending any charge, the law requires a supplier to show that 'all reasonable steps were taken, and all due diligence was exercised to avoid committing the offence'.

This is commonly known as the due diligence defence and this concept is examined in more detail in Chapter 4.

3.9 Summary

Across the EU, prosecutions and other sanctions have been taken against a number of different products. If manufacturers take a responsible attitude, however, they are unlikely to be subject to enforcement action. Manufacturers who do not take all reasonable steps to produce a safe product:

- could have their products withdrawn from the market
- could risk prosecution and suffer bad publicity and loss of business
- could see customers losing confidence in their products
- could face unlimited fines or criminal prosecution that could lead to imprisonment, should a user receive a serious or even a fatal injury.

Chapter 4

'Due diligence'

EU Directives are an integral part of Product Safety law. In common with most Product Safety laws, the LVD aims to provide adequate levels of consumer protection and maintain a fair trading environment for legitimate businesses. As has been indicated in Chapters 2 and 3, the LVD does not contain details of the sanctions and enforcement mechanisms used by individual member states (to explain those processes we used the UK as an example). In this chapter we consider how manufacturers can reduce the risk of any enforcement action being taken against their products. Once again we will use the UK as a model.

The Low Voltage Directive, implemented into UK law by the Electrical Equipment Safety Regulations 1994, is a criminal statute. It is therefore important that manufacturers and importers understand how Product Safety law works if they are to avoid litigation. Once the requirements are understood, it is vital that businesses establish a system to avoid breaking the law – such a system will embrace the concept of *due diligence*.

4.1 How Product Safety law works in the UK

In criminal law, the prosecuting authority has to prove that the person accused of any wrongdoing is guilty beyond reasonable doubt. In many criminal cases, the prosecution also has to prove *mens rea* or guilty knowledge, i.e. they must show an intention to do something wrong.

However, some laws provide an absolute prohibition against doing something. In such cases it does not matter if an offender did not intend to do wrong, or was ignorant of the requirements. If there is sufficient evidence that a particular law has been contravened, that is sufficient to allow a court to convict. The laws governing product safety fall into this category.

This regime can obviously, at times, be unfair. A person might not have been responsible for the offence. It might have been due to an accident, or it might have been committed by somebody else. Add to this the complexity of today's legislation and even the most diligent trader could break the law on a daily basis. Therefore to balance the scales of justice, the UK government has provided various defences and 'let-outs'. This means that the law recognizes the efforts made by reputable businesses to comply with its demands. This system of 'let-outs' includes the defences of *reasonable precautions* and *due diligence*.

To use these defences, a person or business must prove that he took all reasonable steps and exercised all due diligence to avoid committing the offence. These are separate issues, and satisfying *both* is essential – they are not alternatives. In every case, a court of law will have to consider whether the precautions taken by a defendant were sufficient. If he can do so he is entitled to be acquitted.

Whether or not a defence will be successful depends on the circumstances surrounding each case. What amounts to a successful due diligence defence has exercised the minds of many judges over many years and has resulted in a number of appeal cases which in themselves help us to understand more clearly what businesses have to do to avoid prosecution.

4.2 Reasonable precautions and due diligence – the concept

The form of wording for this type of defence is common to most UK consumer protection laws. In the case of the Electrical Equipment Safety Regulations, the defence can be found in the Consumer Protection Act 1987. It requires a business or person to:

> *have taken all reasonable steps or precautions*
> *AND*
> *have exercised all due diligence to avoid committing the offence*

At its simplest, the above requires you to look at the way in which your business operates and then put in place a series of checks to prevent the occurrence of problems. It also requires that you ensure the checks specified by the system are in fact being carried out. If you have a system that nobody knows about, or cares about, the system is useless and a defence plea will fail.

None of the laws that warrant a 'due diligence' defence describes in any detail what systems will actually *provide* a satisfactory defence. To establish this, past decisions of the courts must be examined, and that experience drawn upon. But before any systems can be set up, a clear understanding of the laws applicable to a manufacturer's products is essential. Only through knowledge of the legislated requirements can reasonable safeguards be built.

To assist in this task, there follows some important themes which have emerged from past court decisions on the concepts of reasonable precautions and due diligence:

- Sitting back and doing nothing is unlikely to protect you.
- If a reasonable step or precaution is not taken, any defence is likely to fail.
- Taking reasonable steps is likely to involve setting up a system of control that gives due regard to the risks and the law involved.
- Due diligence means ensuring that your system of checks is both effective and operational, and that you can prove it.
- What is reasonable will depend upon particular circumstances.

Even today, many businesses do not understand the concept of due diligence. Many have expressed surprise that a *defence cannot be made on the premise that*:

- the product has been manufactured for a long time and that there have been no customer complaints with regards to quality or safety
- the manufacturer produces only 'one-off' products and therefore it is impractical to perform testing or evaluation for each one
- the manufacturer produces a wide range of products and is (therefore) too busy to check them, or perhaps some checks are performed, but infrequently
- testing or evaluation is not necessary since the manufacturer is only a sub-contractor and therefore the responsibility lies with the original contractor's designs. In such a case, the sub-contractor builds and tests the product and he is deemed to have the experience to identify any fundamental design faults.

4.3 Reasonable precautions and due diligence – a good practice guide

What follows are some pointers and examples of control techniques. They are not exhaustive or prescriptive, but are intended to provoke thought about protection against legal action.

4.3.1 Assess the risk

Ask yourself what could go wrong in your business that might lead to an appearance in a court of law? An unsafe product, a false claim, a misleading advertisement, unsafe working practices, the list is endless. To assess the risk of such an incident, it is necessary to identify any weak links in the process chain. It is also prudent to know what is happening in that particular sector of industry and also to be aware of how and where your products are being used or marketed.

4.3.2 Establish a plan

Having analysed what could go wrong, reasonable safeguards should be put in place. Have you done everything that the law requires? Are you meeting accepted industry standards? All risks should be controlled by putting in place as many precautions as is deemed necessary. The aim must be either to *eliminate* the possibility of something going wrong (this is unlikely), or to control the risks so that errors *will be detected* and put right before too much damage is done. There is no general formula for creating a due diligence system because each business is different, but in deciding what is necessary and feasible, the following should be considered.

The system which is created must be under the company's control. It is recommended that the responsibilities of all concerned be stated in writing and acknowledged by them.

The precautions and checks to be taken must relate to:

- the size and resources of the company
- the risk posed by the company's products
- other relevant circumstances.

Any system that is devised must be appropriate to the size of the business. The bigger the organization, the greater the legal expectation. The systems created must be appropriate to the consequences of a failure. If sampling or testing is involved *or* appropriate, the number of tests carried out should reflect:

- the cost of getting it wrong
- the volume of product involved
- the complexity of the product
- the cost involved in testing the product
- the degree of confidence in the product
- the size of the business.

Note: *The acquisition of warranties and assurances from suppliers can contribute to a due diligence system.*

4.3.3 Write down the solution

Whatever solution is arrived at, it is essential that the control system be documented so that it can both be followed *and* audited. Without such written evidence it will be very difficult to justify a defence on the grounds of reasonable precautions and due diligence. The documentation will include 'quality control' records, and also supply documents, specifications, training schedules and other documents which show that possible risks have been addressed. The organization's employees should be informed of the actions taken, and training given to those responsible for operating the system of checks.

Documented records should be kept outlining the procedures followed and the checks operated to validate them. These records should be kept safe for future reference. Careful consideration should be given to the length of time records are retained. In some cases there may be a legal obligation in this respect, e.g. most of the CE Marking Directives require that Technical Files be kept for 10 years after the last date of a model's production.

4.3.4 Operate the system

In order to make a defence it has to be shown that a control system had been created, that it worked, and that the system had been operated. This suggests the need to regularly audit the system of checks, and maintain records of the audits. The system should include a procedure for the implementation of corrective actions if things are found to be wrong. All aspects of the system should be regularly reviewed and amended whenever necessary.

4.3.5 Review the system

The system should be regularly reviewed to ensure that it remains effective. No system is 100% foolproof, and over time problems will occur which the system may not address.

The important thing is to show that the system is monitored, and that failures are put right as soon as possible. Once again the importance of written records becomes evident; they can help to identify where and why the problems have occurred, and thus act as a diagnostic device in the monitoring process.

4.4 Reasonable precautions and due diligence – a good practice example

Putting the above into practice will require different actions from different businesses. Creating a system of checks is not easy, but a simple example is given below in an attempt to provide assistance through the process:

- Identify *Safety Critical* areas and prepare a very simple document (perhaps in a flowchart format, see Figure 4.1).

20 Electrical Product Safety

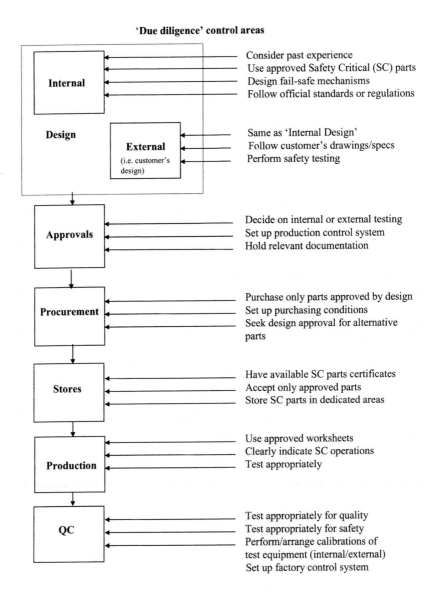

Figure 4.1 'Due dilegence' activities – example flowchart

- Explain the safety related operations within each relevant area by preparing simplified procedures for each one (an example is given in Figure 4.2).
- Collate this information into one document or file (an example is given in Figure 4.3).

'Due diligence'   21

Company Name ....   Issue No. ....

Procedure No. ....   Issue Date..................

Area: Production

- Only trained personnel will perform safety operations

- Worksheet will clearly identify safety operations

- Worksheets will be authorized and dated

- Safety Critical parts will be stored in designated areas

- Every product will be tested for:
  Flash test
  Insulation resistance
  Earth continuity
  before movement to Quality Control area

- Only calibrated equipment will be used for safety testing

- Investigations will be carried out of safety failures and records kept

- All data related to safety testing will be kept for a minimum 10 years

- Etc.

Signed:

Figure 4.2 Operation procedure – example

Display the flowchart on the shop floor in order to make all of the staff aware of the company's duties and actions, and to show how all of the following departments are involved in building a safe and good quality product:

- Design.
- Approvals.
- Procurement.
- Stores.
- Manufacturing.
- Quality Assurance.

```
┌─────────────────────────────────────────────────────────────┐
│  Company Name ..........................   Issue No. ......  │
│                                                               │
│  Procedure No. .........................   Issue Date ......  │
├─────────────────────────────────────────────────────────────┤
│                                                               │
│                     Safety Critical Area                      │
│                            Index                              │
│                                                               │
│   Design ......................   Procedure No. ............  │
│                                                               │
│   Training ....................   Procedure No. ............  │
│                                                               │
│   Approvals ...................   Procedure No. ............  │
│                                                               │
│   Procurement .................   Procedure No. ............  │
│                                                               │
│   Goods in ....................   Procedure No. ............  │
│                                                               │
│   Stores ......................   Procedure No. ............  │
│                                                               │
│   Production ..................   Procedure No. ............  │
│                                                               │
│   Quality control .............   Procedure No. ............  │
│                                                               │
└─────────────────────────────────────────────────────────────┘
```

Figure 4.3 Safety Critical operations index – example

4.5 Conclusions

Creating a system of checks is not easy, and will undoubtedly involve a process of 'trial and error' before an effective system has been established. It is clear, however, that manufacturers *must* take steps to create such a system of checks if they are to avoid the risk of prosecution in the event of a safety related product failure.

Part 2

The practical solution

Chapter 5

Standards – an overview

Standards have been 'around' for a very long time, in fact as early as 2500 BC. The Egyptians were the first to introduce their royal master cubit for noble architectural purposes and the Europeans created weights and measures mainly to prevent fraud in trade. The development of the international standard stemmed from the introduction of the metric system – with the signing of the Treaty of Metre in Paris in 1875 when 18 countries pledged officially to adopt this system.

With the rapid growth and change in the electrical industry during the late nineteenth century, scientists realized that standardization was essential if the industry was to progress. The need for compatibility amongst electrical products and the elimination of technical barriers to international trade were key issues of the time. To meet these needs, the *International Electrotechnical Commission* (IEC) was founded in 1906.

In addition to the IEC, and with the support of the United Nations (UN), a new organization, the *International Standards Organization* (ISO), was born in 1946. This new body was given the responsibility for the co-ordination and unification of industrial standards, particularly those of a mechanical rather than electrotechnical nature.

Today IEC and ISO are the predominant international standards development organizations in the world, and between them they have published many thousands of standards. ISO's membership comprises over 100 National standards bodies, while Presidents of the National Committees of 42 countries make up the IEC's membership.

5.1 Product specific standards

Product specific standards are published in the European Commission's *Official Journal* (OJ) and become national standards. The design and testing of a product to such a standard automatically presumes compliance with the essential requirements of the Low Voltage Directive. In the UK such documents will be published as BS EN, and in Germany they will be published as DIN EN.

In Britain, interest in standardization began at the end of the nineteenth century and the first meeting of the Engineering Standards Committee was held in 1907. After the First World War the name was changed to the British Engineering Standards Committee and finally the name was changed to *British Standards Institution* (BSI) by Royal Charter in 1929 and 1931. Although many BSI standards are based on international publications, a considerable proportion was developed in the United Kingdom as British Standards (BS) and which have in many cases been adopted internationally and in Europe.

Besides the UK, many other European countries have been active in generating standards, in particular Germany and France. Over 4000 electrical standards have been produced by Germany alone since 1896 when VDE (National Certification Body) published its first standard.

5.2 European standards

Since the creation of the single market in Europe in 1993, aiming at the removal of barriers between member states, there has been a strong move towards the elimination of national regulations in favour of the use of common testing and certification methods.

Since the European Economic Community's (EEC) creation in 1957, many organizations have been formed to co-ordinate standards in Common Market countries. One of these is the European Committee for Electrical Standardization (CENELEC – Comité Européen de Normalisation Electrotechnique) which was founded in 1973, its mechanical counterpart (reflecting ISO) is CEN.

CENELEC standards are issued in the following forms:

European Standard (EN) – This is a complete standard, which must be given a national status in all member countries of CENELEC. Conflicting standards must be withdrawn.

Harmonized Document (HD) – This is a preliminary document ahead of publication of the EN. A corresponding national standard must be implemented in all member states. Conflicting standards must be withdrawn.

European Pre-Standard (ENV) – These are prospective standards, they are being used in areas where a high rate of innovation or an urgent need for guidance is required. ENVs are not official standards and although they can be used, existing national standards may still be used until the EN standards become available.

Most European standards are based on IEC documents. Once CENELEC has selected an IEC document as the basic document in a particular area, work by the national committees usually stops until a decision is reached by CENELEC on how to introduce this standard. A procedure known as 'parallel voting' is adopted for the joined development of an IEC and an EN standard together.

By using as the basis for design and testing of a product, a product specific standard which has been published in the European Commission's *Official Journal* (OJ) and which has become a national standard, automatically presumes compliance with the essential requirements of the Low Voltage Directive. In the UK such documents will be published as BS EN, in Germany they will be published as DIN EN.

Harmonized standards covering electrical equipment have been in existence for a long time. Going through the already published and regularly updated BSI standards catalogue, one may be surprised to see how many British, European (Harmonized) and International standards exist covering most product ranges.

The list of published, *Harmonized* standards for compliance with the LVD presently covers:

- Components.
- Household equipment.
- Information technology equipment.
- Lasers.
- Installations.
- Lighting.
- TV and Radio.
- Tools.
- Welding equipment.
- Machines.
- Etc.

Some of the standards which cover a whole range of products are:

- EN 60335 – Safety requirements for household and similar electrical appliances. Its scope covers large and small kitchen appliances, vacuum cleaners and (under the 'similar' category) commercial catering and vending machines.
- EN 60065 – Safety requirements for mains operated electronic and related apparatus for household and similar general use. Its scope covers TVs, video recorders, hi-fi and personal stereos, etc. A summary of the requirements of this standard can be found in Appendix 16.

These two standards cover most products found at home, with a few exceptions for specific product areas such as electric blankets (EN 60967) and handheld power tools (EN 50144).

Other widely applied standards are:

- EN 61010 – Safety requirements for electrical equipment for measuring, control and laboratory use.

28 Electrical Product Safety

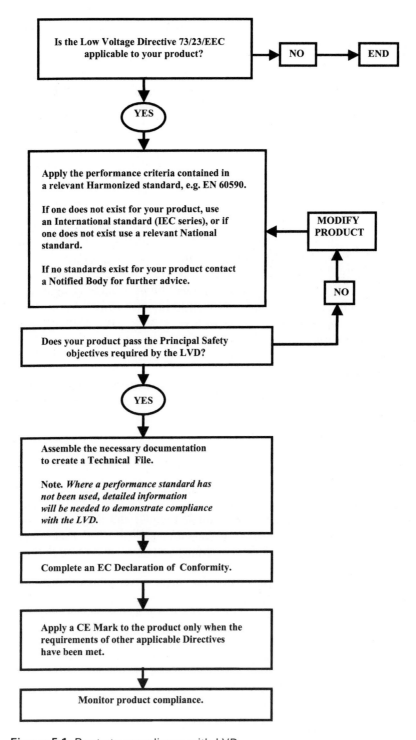

Figure 5.1 Route to compliance with LVD

- EN 60950 – Safety requirements for information technology equipment including electrical business equipment.

5.3 When product specific standards do not exist

For products which are new or innovative and lie outside the scope of an existing suitable standard, the manufacturer should arrange for the product to be tested by a Notified Body. 'Good engineering practices in safety matters' will have to be established and detailed explanations of measures taken to ensure compliance will need to be listed in the Technical File.

A sensible approach would be to determine a standard which most closely relates to the product and use this as a basis. Excluding unnecessary tests while adding others creates an appropriate programme of tests which results in the complete evaluation of the product.

A simplified flowchart as given in Figure 5.1 should help the manufacturer through the process of identifying a suitable approach towards gaining confidence before CE Marking his product.

The majority of national standards in Europe are voluntary and manufacturers don't have to follow them as long as the essential requirements of the Directive are met. However, some standards – particularly those related to safety – have become mandatory by legislation at national level. An example of mandatory legislation is the UK Plugs and Socket Regulation (1987) which requires compliance with BS 1363.

5.4 Summary

Harmonized standards exist for most product ranges. A designer should make a point of familiarizing himself with a product's specific standard and design according to its requirements.

BSI issues catalogues of Harmonized, International and other standards that can be used to show compliance with the LVD, some of these are listed in Appendix 2.

When Harmonized standards do not exist, the manufacturer can use alternatives as shown in Figure 5.1.

Chapter 6

Testing for safety

Manufacturers of electrical equipment who have not previously been involved with safety testing, but suddenly become aware of the Low Voltage Directive's requirements, often ask themselves:

Do I really need to test?
Do I need a Certification mark?
How do I get it?
Where can I test and how long will it take?
Which standards must I test against?
How much will it cost?
How do I ensure compliance?

It is surprising to hear these questions still being asked even though the LVD has been in place since 1973! In the next few paragraphs an attempt will be made to provide answers to all of these questions, and thus provide advice to help an organization decide the best course of action for its particular product or product range.

6.1 To test or not to test?

Without any doubt, testing to a recognized specification is the key element for establishing compliance with the technical requirements of the Directive. In some cases, the customer's purchasing policy may allow no option, i.e. 'we will only buy products tested and certified to ... by ...'.

Gaining independent certification can help a company avoid civil and criminal actions. In the case of a serious incident in the market where a product has caused injury – perhaps completely destroying itself in the process the ability to demonstrate that it (the marque) had been designed safely and had been independently assessed can be a persuasive and sufficient defence.

Testing a representative sample from every product range would be the ideal situation for manufacturers of electrical products. However, few manufacturers have the financial resources or the expertise to perform compliance tests – either 'in-house' or at a test laboratory – every time they add a new product to their range. Particularly small companies have to rely on pre-compliance tests, self-assessment or technical justifications in order to justify not testing before release of their products to the market.

If you are one of those suppliers who rely on product self-assessment without performing any testing – your justification being based on designing safely with a good understanding of safety principles – you must appreciate that you will not have the same level of protection that complete testing to a Harmonized standard by an independent accredited test laboratory provides.

It might help you though to demonstrate due diligence and give yourself some confidence that the product is safe – thus meeting the 'essential requirements'. In any case, it is definitely better than doing nothing.

6.2 Do I need a Certification mark?

It is not necessary to be independently certified by a National Certification Body (NCB) in order to affix the CE Mark to a product, i.e. no other Certification mark is needed.

Until a few years ago, a nationally recognized Certification mark was a mandatory requirement in most European countries for the majority of electrical products – in particular for household appliances. In the UK, both BSI (British Standards Institution) and BEAB (British Electrotechnical Approvals Board) are the National Certification Bodies for the purposes of the CENELEC certification scheme (the CENELEC Certification Agreement or CCA) for the Low Voltage Directive. The CCA scheme allows the simplified approval of products by all CCA scheme members after successful testing has been completed by any one member of the scheme.

Certification for most products was never compulsory in the UK, but that was not the case for some other countries, e.g. approval before import into Scandinavia and Switzerland was necessary.

As explained above, in most cases the market (the *customer*) will dictate whether or not a Certification mark is required – large retailers still insist on it.

6.3 How do I get a Certification mark?

Certification marks are offered by all NCBs across Europe. Manufacturers may approach any of the following:

BSI or BEAB in the UK
VDE in Germany
NEMKO in Norway (test laboratories also in the UK)
FIMKO in Finland
SEMKO in Sweden
KEMA in the Netherlands
DEMKO in Denmark

and many others.

From experience it makes sense to deal with a local NCB – it can help to reduce the costs associated with both the transportation of samples, and travel to the test laboratory in cases of problems. Communications are also greatly simplified through the use of a common language and national as opposed to international telephone and fax services.

When making an application, the Certification Agency will usually request:

- a completed application form – listing Safety Critical parts and their approval status
- a sample product
- spare parts – needed to replace components during fault testing
- copies of certificates of approved parts/mouldings, etc.
- a set of circuit diagrams
- any relevant test reports
- a copy of the instructions of use.

Testing and approval can then proceed. Upon successful testing a test report will be issued by the laboratory and sent to the applicant together with an approval Certificate.

Certification Agencies that carry out tests in accordance with the Low Voltage Directive will also (necessarily) operate a factory approval scheme whereby the manufacturing premises will be audited (normally annually) against the requirements of CCA document 201.

Certification by an NCB is also possible without the need for a factory audit, but then the NCB's approval mark cannot be affixed to the product.

6.4 Where to test and how long will it take?

As an alternative to product testing by a National Certification Body, manufacturers also have the option (in the UK) to apply for testing at an independent UKAS (United Kingdom Accreditation Service) accredited test laboratory. The testing will be performed to a Harmonized standard (if available) and the issued test report can be used by the manufacturer as part of his technical documentation file (Technical File) in order to provide evidence of compliance with the LVD. Choosing this route, however, precludes the right to affix a Certification mark to the product.

Testing for safety

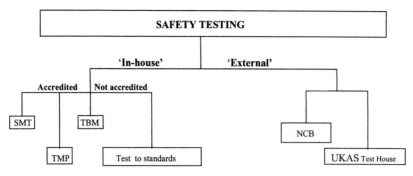

Figure 6.1 Options for testing

ERA, TRL, TÜV Product Services and other laboratories in the UK offer this service.

Until a few years ago, testing by a small number of Test Houses was the only way to obtain confidence that products complied with relevant safety standards. This is now changing through the introduction of various 'in-house' testing schemes offered by an NCB such as BEAB, DEMKO, UL, etc.

Test times vary depending on the standard used and the complexity of the product. It can take as little as two weeks or as long as six to eight weeks, sometimes longer depending on the 'queue' at the Test House. Time for testing may also vary with seasonality – holiday periods in particular should be avoided.

SMT (Supervised Manufacturer's Testing scheme) – Manufacturer's test laboratory accredited by an NCB to EN 45001. Testing is performed by the manufacturer's personnel under the periodic supervision of the NCB.

TMP (Testing at the Manufacturer's Premises scheme) – Manufacturer's test laboratory accredited by an NCB to EN 45001. Testing is performed at the manufacturer's laboratory by the NCB personnel.

TBM (Testing By the Manufacturer) – Manufacturer's test laboratory accredited by an NCB to EN 45001. Testing is performed at the manufacturer's laboratory by the manufacturer and confirmed by the NCB at their own laboratory. After 1 year a TBM facility is normally accepted without further confirmation by the NCB.

NCB – Testing is performed either by the NCB's own laboratories or by NCB authorized laboratories (i.e. sub-contractors).

UKAS approved Test House – These are independent accredited test laboratories competent to test to Harmonized standards, e.g. ERA, TRL, etc.

As an alternative to the above, a manufacturer may, if he so wishes, perform testing in his own non-accredited laboratory, provided that he employs trained and competent personnel and has the correct (and properly calibrated) equipment. In this case the manufacturer may create his own test reports and self-certify his products.

If a manufacturer, for whatever reason, does not perform full in-house or external testing, a very thorough product self-evaluation (with some minimal testing) may be acceptable. In the following chapters we will attempt to explain how to perform this evaluation and how to complete a report.

6.5 Which standards to test against?

Choosing a standard for safety testing requires careful consideration. This is covered in some detail in Chapter 5.

6.6 How much will it cost?

Cost will depend on the product and the applicable standard. In order to ensure that you get the best deal, follow the few simple steps given below.

Prepare an information pack containing:

- a summary of the product and its performance
- a block/interconnecting diagram showing relevant circuitry
- details of the standards against which you wish to have the product tested
- details of the approval status of critical components.

Then send it to a number of NCBs or Test Houses, requesting quotations.

You will be surprised at the variation in quoted costs – anything from £800 to £3500 for the same product is not untypical!

With the need to perform third party testing now removed, NCBs and Test Houses have lost their monopoly. Some have become commercial organizations, and some are struggling to survive – but all have become more flexible and less costly! *But beware – some very low cost tests do not always offer what they promise.*

Some very attractive offers are now available including the 'One Stop Shop' where testing for Electrical Safety, Electromagnetic Compatibility (EMC) *and* Machinery Regulations is all done using one sample at one location.

6.7 How do I ensure compliance?

All manufacturers of electrical equipment, irrespective of their scale of production, have to meet the requirements of the LVD and be able to demonstrate 'due diligence'. Actions which may be classed as reasonable and acceptable for one business, might be insufficient for another. To help clarify specific responsibilities, a few suggestions are made below, but these should only be viewed as *the minimum requirement*.

(i) The 'ideal situation' – although in practice this can only be practised by large-scale manufacturers producing a variety of products:

- Samples from every new product range are tested (either in-house or externally), using the NCB route.
- Implement a comprehensive Production Control system including final basic electrical testing of each production unit (may include earth path continuity, dielectric strength test, insulation resistance and leakage current).
- Operate a Change Control system (changes to the original design that affect safety are be controlled and authorized).
- Operate the factory in compliance with the NCB's factory requirements.
- Arrange regular product retesting (if the same product will be manufactured for a long time).

(ii) Not all manufacturers are able to conform to all of these requirements, however; smaller organizations tend to:

- Test quite irregularly (in-house or externally), or not test at all.
- Operate a very basic production control system.
- Not have a change control system in place.
- Not obtain external factory approval.

For those companies which fall under this category and do not wish or cannot afford to deviate from their present system, it is advised that as a *minimum* requirement, they should:

- Perform some product testing or comprehensive evaluation (with some basic testing included).
- Set up a basic factory control system (paying particular attention to the final tests before despatch), and maintain tight control over the use of safety critical components.
- Set up a Change Control system for safety critical components (at least one person from the Design, Production or Quality Control departments should be designated to carry out this task).
- Identify product safety critical areas within the plant and introduce a safety awareness programme.

From the information provided above, the reader might draw the conclusion that the use of an NCB and testing according to the standards is the only way to ensure compliance with the safety objectives of the LVD. This is not strictly true – in fact, only a small proportion of manufacturers of products under the scope of the LVD follow this route. The alternatives outlined in the following chapter may be found useful.

6.8 Summary

It is a fact that testing to ensure conformity with all the applicable standards and regulations will not *guarantee* that products will be safe in all circumstances (thereby automatically freeing the manufacturer from risk of prosecution and liability damages).

A sensible test programme, however, together with certification from a recognized National Certification Body, and an adequate Production Control system, will provide a first line defence should the safety of a product be called into question.

Chapter 7

Self-evaluation – route to compliance

The cost of complying with so many EU Directives has undoubtedly placed a heavy burden on manufacturing industry, particularly small business. From the authors' experiences, this is cause for concern for many small organizations and many are looking for alternative ways to reduce this expense.

In order to help small businesses meet the requirements of the LVD without incurring the additional costs incurred by external product evaluation, we have prepared some brief guidelines that could be of use to small-scale manufacturers of electrical products. It must be pointed out, however, that although what follows is considered to be best practice, it holds no legal status and it must be read in conjunction with the Low Voltage Directive.

Let us now consider the scenario where you, the reader of this guide, are the designer of a new product. You have some experience in using safety standards and you are reasonably familiar with safety principles. By following the step-by-step approach described below you should be able to design and self-evaluate/test your product, and be confident of meeting the safety objectives contained in the LVD.

7.1 Step 1 – Finding an applicable standard/regulation

As a first step, you need to establish which safety standard or regulation is applicable to your proposed design. You should start by preparing an information pack that includes as much information about the product as possible. As a minimum it should contain:

- the product's concept
- where it is intended to be used

- its power supply requirements
- potential market(s)
- size/weight, etc.

This information could then be submitted to the local National Certification Body or a Notified Authority/Test House with a request for advice on:

- whether the equipment is included in the scope of the LVD
- the most applicable safety standard or regulation (and the latest issue).

Many organizations will not charge for such a service.

7.2 Step 2 – Understand the fundamental safety requirements

If a standard or regulation is applicable to your products, you should become familiar with its fundamental safety requirements. Obtain a copy (see Note below) and start by studying this document. For the novice this will be a slow, lengthy process and certainly confusing at first. To help you further, it might be advisable to create (in order) lists of the following:

- the main core requirements
- minor requirements (preferably clause by clause) – see Chapter 8 and Appendix 6
- equipment necessary to confirm your design upon completion.

If at any point you feel unsure about the tests or evaluations listed in the standards, seek clarification from your NCB or a Test House.

Note: *Safety standards and regulations can be obtained from BSI in the UK (for address see 'Acknowledgements' and Appendix 3); for other EU countries contact your local NCB (see Appendix 4). Make sure that you request details of the latest version and any relevant Amendments.*

7.3 Step 3 – During the design stages

Once you have read and understood the safety standard or regulation, you can now start your design. Usually the design is completed in modules, and during these early design stages consideration should be given to each module's design parameters which are critical to the final product's safe operation under normal operating conditions and under fault conditions, i.e.:

- overcurrent protection devices
- component spacing (for insulation of flammability)
- primary (high voltage)/secondary (low voltage) circuit separation

- leakage/touch currents
- insulation distance and thickness (refer to Chapter 10)
- use of approved components (see Step 5)
- etc.

These are important parameters and should be considered while the new product is still on the 'drawing board'. Retrospective design changes can be very expensive and possibly impractical to apply, so early stage considerations are paramount in saving time, effort and cost.

7.4 Step 4 – Test equipment considerations

Although you may think it is too early to consider what equipment you will need to verify your prototype, you may find that some specialized equipment suppliers have a long lead time for purchase or hire.

Consider what equipment you will need to carry out:

- mechanical tests (you may need specialized instruments)
- temperature measurements
- voltage, current and power measurements
- leakage currents
- flash (dielectric strength), insulation resistance and earth continuity
- etc.

For test routine guidelines, see Chapter 10, and for equipment and suppliers of test instruments, refer to Appendices 5, 11 and 18. You may find that most of the equipment you will need is already available in your Design, Quality Control or Production areas – but before you decide to use them for compliance testing, ensure they are calibrated!

For specialized tests, you may of course need to purchase or hire some additional instruments or fabricate 'similar' tools in-house.

7.5 Step 5 – Decision on Safety Critical parts

During the design stages, you will need to consider which parts and components are important to guarantee continuing safe operation under normal operating and fault conditions (refer to Chapter 9 for details on Safety Critical components and their approval requirements).

You will need to list which components or parts are considered Safety Critical (SC) and start the selection process at this stage.

In order to achieve confidence in meeting the standards and designing a safe product, you should select only SC parts or components with external electrical safety approval, or those which meet the flammability requirements as specified in the standards.

Deciding on SC components is very critical, and in parallel with obtaining their operational specifications from the supplier (necessary if the product is to perform as intended), you should obtain confirmation of their safety approval status before you decide to use them.

7.6 Step 6 – Prototype testing

When you have a prototype available, you should consider performing initial confirmation tests.

In an ideal situation, there should be an area specifically allocated for safety testing with its own dedicated instruments and equipment, but this is not always possible. Having collected the necessary tools and equipment as explained above, you will need to find a 'quiet corner' to start your evaluation.

Begin by looking at fault testing and other tests that are affected by the circuit design; external construction tests could follow later. Using the checklist created earlier (see Step 3), perform as much testing as possible and confirm compliance to the basic safety requirements before the design moves to the next stage.

7.7 Step 7 – Creating an Evaluation Report

Details on compiling a product Evaluation Report and an example of such a document can be found in Chapter 11.

7.8 Step 8 – Testing the final product

If the product you have designed is one of a range of similar products, select the most complicated of the series and start your evaluation.

Tests should be performed in the order that you feel is most convenient or appropriate. For example, you should perform some humidity treatment (possibly in an environmental chamber) if the product is likely to be used in a humid environment, and immediately after the humidity treatment you should perform the dielectric strength test.

After the first two or three evaluations you will settle into a routine and be able to perform testing in the most convenient manner. During the testing/evaluation, you should carefully consider the aim of the tests – some of these could be:

- to identify all risks associated with your product's operation (this includes electrical and mechanical issues)
- to perform as many tests and evaluations as are necessary to ensure that all the risks have been assessed
- to gain the confidence that your product has been evaluated correctly
- to confirm that the product has passed the required tests or evaluations relevant to the particular standard or regulation
- to confirm the product has passed the principal safety objectives

- to confirm that it is 'safe' to be released to the end user.

Do not forget to perform the important tests of dielectric strength, insulation resistance, leakage current and earth continuity (if applicable).

7.9 Step 9 – The final act

Having evaluated your product and being satisfied that it meets the 'principal safety objectives', you need to:

- prepare a Technical File (see Chapter 12)
- prepare an EC Declaration of Conformity (see Chapter 13)
- affix the CE Marking on your product.

7.10 Summary

The principal safety objectives are defined in the Low Voltage Directive. They require products to be built in such a way that electrical equipment is:

- safe
- constructed in accordance with good engineering practice
- designed to met the safety requirements contained in the LVD.

Self-evaluation may be the only possible way for a small organization to meet the LVD and supply its products across the EU. This is recognized and accepted by the enforcement authorities. If you rely on self-certification you must take all reasonable precautions to supply a safe product, and you should therefore:

- confirm that your product is under the scope of the LVD
- search for an applicable (or closest relevant) standard for the goods produced
- become familiar with the relevant standard or regulations
- understand the standard's requirements and the safety principles listed therein
- ensure that the design work is carried out with due regard to the standard
- decide what tests can be performed 'in-house' and obtain the necessary equipment;
- perform the tests and confirm compliance
- complete a self-assessment (evaluation) report and keep it with the Technical File.

You can also help yourself further and reduce your approval costs by:

- applying only relevant standards or regulations
- evaluating your product for the environment of its intended use

- taking the user's expectations into account
- providing sufficient warnings and detailed lists of limitations of use
- comparing it with similar products externally tested previously
- using your 'in-house' expertise.

Finally, act in the spirit of the Low Voltage Directive – you should do everything practicable to remove any doubt about the safety of your product.

Chapter 8

Fundamentals of product safety

Clear understanding of the basic safety principles is an essential tool for any good design engineer. If an appropriate standard exists, there is no better substitute for knowing exactly what the requirements are and what criteria have to be met. If, however, no relevant standard is available, then the likely conditions and environment of use must be considered.

There are hundreds of safety standards available to choose from, each one having particular requirements unique to itself, as it would be impossible to include all the requirements, test methods and acceptance criteria in one single document. In this guide an attempt has been made to give the reader an understanding of what the *basic* considerations should be when designing for safety. Some of these considerations may be easily overlooked during design when the focus is on performance, customer expectations and cost. Consideration of safety requirements is essential as retrospective countermeasures for non-compliance could be very costly.

As explained earlier, although most safety principles are common to most of the safety standards, the requirements will not be the same for every product. You therefore need to carefully consider what requirements pertain to your product and apply the relatively simple principles set out in this guide. Designing for safety may then become almost instinctive – perhaps to the point of automatically achieving compliance with relevant safety laws and standards.

Electrical equipment should operate safely under both normal and fault conditions. This is a clear requirement of most safety standards and of course of the Low Voltage Directive. Electrical faults can result in equipment overheating and the possibility of a fire within the equipment which may even spread outside its enclosure. All necessary precautions have to be taken to minimize such risks.

Most standards specify the use of suitably flammability rated enclosure material, and some also specify similar requirements for internal plastics and wiring. Selecting suitably rated material at the design stage will minimize costs and time by obviating the need for rework or redesign in the event of the product failing to meet the standard when tested in a test laboratory.

For information technology equipment, EN 60950 specifies the need to use suitable flame retardant material for most components – with the exception of small parts. In addition to the risk of fire, consideration has to be given to the temperature rise on internal and external components which might result in injury to the user. During testing (when the equipment under test is operated continuously under normal operating conditions in a fully loaded condition) temperature rises are measured, but only after a steady state is reached.

Tests for temperature rises are also performed under fault conditions. An acceptable result from these tests would include the operation of the equipment within specified temperature limits, the non-operation of thermal cut-outs, and sealing compounds not melting and flowing out.

Some standards (such as EN 60950) have reduced fault testing requirements as they quote requirements for suitable flammability ratings, while others (such as EN 60065) require more intensive testing and have limits for temperature rise of components under fault conditions. This reflects differences in the type of product and the environment in which it is intended to be used.

Some of the most basic principles considered necessary for product safety are listed in Tables 8.1 and 8.2.

In addition to the above, there are some other hazards which are dealt with by only some specific standards, these are:

- *Explosion* – from batteries.
- *Ultraviolet radiation* – from copiers and fax machines.

Table 8.1 Summary of basic safety principles

Safety principle	Compliance can be achieved by using
Protection against electric shock	Appropriate barriers (i.e. enclosure) and insulation, construction of earth path
Protection against fire	• Low flammability internal components • Minimizing risk of ignition • Fire enclosures
Protection against mechanical hazards	Suitable guards and interlocks if necessary
Protection against temperature rises	Adequate design safeguards including overcurrent protection and thermal cut-outs if needed
Protection from energy hazards	Barriers or suitable discharge circuits

Fundamentals of product safety 45

Table 8.2 Summary of some special safety hazards

Special requirements	Compliance achieved by
Protection against radiation hazards (e.g. laser, X-ray)	Employing suitable guarding and enclosures
Protection against chemical hazards	Using suitable sealed cases
Protection against implosion hazards (CRT)	Using suitably approved components

- *Microwaves* – from medical equipment and microwave ovens.
- *Noise* – from vibration machines and ultrasonic test equipment.

Generally, requirements only apply to the hazards as they relate to the exposure to the user; service people are considered to be more knowledgeable in this respect. However, if the service person is expected to access areas where he is likely to be exposed to electric shock, temperature or mechanical risks, then adequate protection must be built into the product.

Definitions:

- *User-accessible area* is any part of the product which can be accessed without the use of a tool.
- *Service-access area* is any part that requires a tool for access. A tool can be a screwdriver or a key, for example.

Details of what are and how to avoid risks from some of the main hazard areas are given in the following sections.

8.1 Protection against electric shock

Risk from electric shock is defined as likely to be present if the operator is exposed to voltages exceeding certain levels.

The limits for acceptable voltages and currents will vary between standards and products – this point will therefore have to be confirmed for the reader's own particular product.

In general, protection against electric shock can be achieved by:

- Using a suitable *enclosure* – this will provide the barrier between the user and the hazardous voltage areas inside the equipment.
- Employing an adequate level of protection within the cabinet by the use of either a *protective earth* (Class I protection) – commonly used where a metal cabinet is employed or (when a non-metallic cabinet is used) through the use of *double insulation* (Class II protection). In this case operator protection is achieved by using double or reinforced insulation, so providing a barrier between high voltage areas and components connected directly to user accessible parts. Often Class

II protection is provided in handheld domestic products through the use of a plastic moulding around all electrical parts.
- Using *insulated parts* where high voltages are directly accessible by the user, e.g. through the power cord or plug.
- Using suitable *guards* to cover openings, so as to avoid access to dangerous voltages by hand or by other objects.
- *Segregation* of conductive fluids and moving parts from electrical parts.

8.2 Protection against fire

A source of combustion and some fuel could result in a fire. Polymeric materials (such as plastics) are common nowadays in most electrical and electronic products. They provide an ideal fuel placed adjacent to possible sources of combustion (parts which can arc or heat unacceptably). Fire risk is a main concern of most new or revised standards, e.g. IEC 65 standard 6th edition (recently published as EN 60065:1998) has included many additional flammability requirements. Protection against fire can be achieved by:

- Using an *enclosure*, with suitable flammability rating.
- Using *internal wiring* and *connectors* with suitable flammability rating.
- Using *printed circuit boards* with suitable flammability rating.
- Minimizing *temperature* rises, thus minimizing the possibility of combustion.
- Maximizing *spacing* around high voltage components to prevent arcing and so minimize the risk of fire starting.
- Providing *overcurrent protection* and thermal cut-outs.

8.3 Protection against mechanical hazards

Risks of mechanically caused injuries can be associated with any of the following:

- gears
- fans
- blades
- linkages
- instability
- sharp objects

Protection against mechanical hazards can be achieved with any of the following:

- Using a suitable *enclosure*, acting as a barrier between the operator and the hazard.
- Where *openings* are necessary (e.g. for ventilation), precautions need be

taken to *prevent access*. The dimensions of openings (length and width) are determined either:
(a) to prevent access by the operator (finger, hand or other bodily part) or,
(b) to prevent the entry of foreign bodies, e.g. conductive dust, coins, paper, paperclips, etc. (must be of a much smaller size than those in (a) above).
- Providing a *safety switch* or an *interlock* to prevent operation and exposure to dangerous parts if protective covers or guards are removed.
- Providing sufficient *labelling* warning the operator of the potential hazard.

8.4 Protection against temperature rises

Excessive rise in temperatures can result in damage to the insulation used to isolate hazardous parts. Additionally, high temperatures generated within the equipment may be transferred to user-accessible metallic parts and cause injury to the operator. Magnetic parts (transformers, relays, inductors, etc.), capacitors, switches, PWBs, isolators and fuses must be measured for such rises.

Protection against temperature hazards can be achieved by:

- Using *suitable components*, i.e. of a correct temperature rating.
- Designing *thermal cut-offs* in the circuitry of the equipment.
- Using a suitable *enclosure*, acting as a barrier between the operator and the hazard.
- Providing sufficient *spacing* between high temperature components and metal user-accessible parts.
- Providing *overcurrent protection*, e.g. fuses or circuit breakers.

8.5 Protection against energy hazards

A low voltage component such as a battery or a capacitor can hold large amounts of energy. A short-circuit or an earth contact inadvertently applied to a bare live part in a high energy circuit at any voltage can cause metal to melt with arcing and burning. High currents can circulate from low voltage circuits also – covering or other protection of low voltage parts may therefore be essential. Equipment that employs large capacitors across the supply poles should employ discharge paths so that energy remaining in the capacitors after equipment switch off can be safely discharged. This may be necessary to prevent shock hazard from parts of the equipment such as the pins of the mains plug.

8.6 Protection against dangerous chemicals

Most safety standards are vague in this area and designers must rely on taking sensible precautions and give warning by labelling of the equipment and provide clear instructions for use.

Flammable liquids or vapour should be kept away from any parts which could cause ignition by arcing in either normal or fault condition. It may be necessary to use *separate compartments* for liquids and for electrical parts, and precautions should be taken to avoid chemical spillage.

8.7 Other hazards

Protection against other possible hazards generated due to *radiation* (e.g. laser, X-ray), *implosion* (CRT), *radio frequency, ultraviolet, ionizing radiation* and *acoustic vibration* should also be considered. Precautions should be taken to ensure safe operation of the product, and the user must be protected from all of the above hazards. This can be achieved by appropriate (and sometimes complicated) design. Individual components are usually tested to specific standards and approved by dedicated and specialized laboratories.

8.8 Summary

Basic safety principles cover hazards from *electric shock, heat, energy, radiation, chemical, fire* and *mechanical sources*. By taking into account these considerations, a design engineer will be able to design a product that will meet the core requirements of most electrical standards.

Chapter 9

Choosing Safety Critical components

Not all components used in an electrical product have safety implications and need to be classed as *Safety Critical*. As a general rule, components whose failure might result in the product becoming unsafe should be of an approved type. The designer needs to clearly identify parts which fall into this category if he is to design a safe product and thereby reduce unnecessary expense in retesting and rework.

To help the reader with the identification of Safety Critical components a list of such components and their function has been compiled and is given later in this chapter (Section 9.4).

A product submitted for safety approval testing may fail to meet the requirements of a particular standard if the components used in its construction are of poor quality and do not comply with their own relevant safety standard. Problems can often be avoided if approved components are used.

As an alternative to using approved components, the test laboratory may be requested to test specific components as part of the product approval. This is a practical and common procedure, but is only applicable to certain components, e.g. resistors, capacitors and transformers, etc. For most other components (such as fuses, relays, AC inlets, power cords, switches, etc.) this is time consuming and very costly. In this case the use of approved components is a much better option.

There is a range of standards which specifically apply to components such as fuses, transformers, resistors, capacitors, control devices, etc.; for some of these, third party testing is often required. Approved components are mainly intended for incorporation into final products and, in addition to safety considerations, they may also have been tested for functional reliability (e.g. mains switches), as well as their response to environmental stress and possible failure modes.

Some components falling within the scope of the LVD will carry CE Marking even though they are intended to be incorporated into larger systems (e.g. power supplies). Many manufacturers of components which are approved by a third party and are compliant with one or more standards claim that their products are 'CE approved'. Such claims cause confusion as there is no such thing as 'CE approval', the only valid 'CE' recognition is an unqualified 'CE' logo (i.e. the CE Mark itself) – third party approval of components falling under the LVD is not a mandatory requirement.

Some safety standards indicate that where safety is involved, components used shall comply with the requirements of the standard's relevant clauses or those of another relevant IEC standard. If IEC compliant components are used, the designer must ensure that:

- the declared standard is relevant to component type
- the component used is correct for the application and environment of use
- it has the correct rating (voltage, current, temperature, etc.).

9.1 Component approval/selection – options

Not all safety critical components are supplied with third party approval – the vendor may choose to offer them either with or without this approval, leaving it up to the final product's manufacturer to decide which parts to buy.

When a component is placed in the market the vendor has the choice:

- of testing 'in-house' according to a recognized EN or IEC standard relevant to the component, and self-declare compliance
- of obtaining third party approval by a National Certification Body to a recognized EN or IEC standard, and following Certification, marking the component with that Certification Body's approval mark
- of supplying the part without approval.

The finished product designer, having identified the Safety Critical components in the new design, also has options and may:

(a) choose 'off-the-shelf' approved components
(b) choose non-approved components but submit them for testing and approval as part of the product's overall testing and certification.

In most cases option (a) is preferable but not always practical, as the designer's decision might be influenced by the increased purchase price of the approved component, the availability of an approved part to the correct specification, the correct mounting configuration, etc.

9.2 Using approved components

Using approved components may seem to be a good solution to problem-free testing and product approval, but the manufacturer may need to take some precautions:

- Components claiming approval to EN standards should be checked to confirm if the approval is for safety or EMC or both.
- The standards to which components are approved are relevant to the intended application.
- The component's parameters, such as voltage, current rating and temperature limitations, will not be exceeded, and the mounting configuration is suitable for the intended application.
- The approval mark is applicable for the countries in which the product will be sold. For acceptance in Europe, the approvals held should include a European approval. UL (USA) and CSA (Canada) approved components may not have been assessed to European (EN) or internationally recognized standards (IEC). Similarly, if the components have EN approval but are destined for the North American market, their approval may not be accepted when submitted for testing. Components which have been approved to 'old' British Standards (BS) may have been superseded by ENs and may be invalid.
- The approval certificate specifies the part intended to be used, as some component variants are not always listed in the certificate.
- Declarations of Conformity, for components carrying the CE Mark only, must refer to a particular standard or regulation.

In addition to the above, it is advisable to confirm the current validity of any approval certificates – the validity may vary from two to five years (depending on the Certification Body). Some components, such as fuses or mains plugs, are manufactured under a special scheme where regular inspections of the manufacturing premises are carried out by the Certification Body (e.g. ASTA or BSI).

The designer should make a list of Safety Critical components used in the product, and reference it against a list of all the applicable standards (EN, BS or IEC). Only through a close familiarization with the scope and requirements of these standards can he be sure that the correct approved components have been chosen.

Obtaining approval certificates from some suppliers might prove difficult, but reference to the supplier's technical documentation may help to identify the standards against which they have been approved. If possible, a library of approval certificates should be created for quick reference. BSI catalogues provide lists of approved components and their applicable standard, and also provide a cross-reference between equivalent standards (BS, EN, ISO and IEC).

9.3 Approval marks

Most components are approved by a third party (usually a National Certification Body), and will be marked either with that NCB's approval mark, or with the applicable standard's reference number.

Some BSI approved components are not marked (the supplier should be able to provide details of the approvals) but some carry the 'Kite' mark (usually fuses and plugs) which indicates that the component manufacturer has a QA system in place.

Power cords and insulated cables will often be marked with the ⟨HAR⟩ mark. These parts are covered by the European CENELEC Harmonized standard, implying that the cables have been type tested, normally in the country of manufacture. It also implies that the factory is regularly inspected by the Certification Agency.

The mark may appear in more than one format, depending on the approval agency. Marks such as CEBEC ⟨HAR⟩, ⟨DEMKO⟩, etc. are acceptable. Some European approval marks which may be found on components are listed below:

VDE – Germany

DEMKO – Denmark

NEMKO – Norway

SEMKO – Sweden

FIMKO – Finland

BASEC – British Approval Service for Electrical Cables, UK

BSI – British Standards Institution, UK

ASTA – the Association of Short Circuit Testing Authorities, UK

KEMA – Netherlands

SEV – Switzerland

CEBEC – Belgium

OVE – Austria

NF – France

9.4 Safety Critical components – examples

Table 9.1 Examples of Safety Critical components

Component	Comment	Possible standard
AC inlets and outlets	Provide protection against electric shock.	EN 60320
Circuit breakers	Under fault condition interrupt the power supply and provide protection against electric shock and fire.	EN 60898 EN 60934 EN 61008-1/-2-1 EN 61009-1/-2-1 HD 536 S1/537-S1
Couplers	Provide protection against electric shock.	EN 60320-1-series
Cathode ray tubes	Provide protection against the effects of implosion or mechanical impact, particularly due to glass fragments.	IEC 65/ EN 60065
Enclosures (other than metallic)	Provide protection against fire and electric shock and access to mechanical hazards. They must have adequate strength, have minimum thickness and be made of slow burning material. The base material must be approved.	Product specific standard
Degaussing coil	Connects between primary and secondary circuit, it must be adequately insulated.	Product specific standard
Fuse/Fuse holder	Provides protection against fault conditions preventing the product from unsafe operation. Its rating characteristics (voltage, current and operation) must be suitable for the application.	Miniature fuses: EN 60127-1/2/3/6 Low voltage fuses: EN 60269-1/2/3
HV components (HV leads, flyback transformers, etc.)	High voltage parts and cables carrying more than a typical value of 4 kV must not support a self-sustaining flame for more than a few seconds (typically 30 seconds).	Product specific standard
Isolating resistors	Provide a discharge path between HV (primary) and LV (secondary) circuits. Distance from adjacent parts must be controlled.	Product specific standard
Isolating capacitor 'Y'	Used for the filtering of transients or reduction of earth currents between primary and secondary circuits.	IEC 384-14, 2nd Edition Other capacitors: EN 60143 EN 60252 EN 60931-1/2 EN 61048 HD 207 S1/597 S1

(*continued*)

Table 9.1 Examples of Safety Critical components (*cont.*)

Component	Comment	Possible standard
Mains isolating capacitor 'X2'	Used as a filter capacitor between 'live' and 'neutral'.	IEC 384-14, 2nd Edition
IC protectors/ fusible resistor	Operates in a similar way to a fuse, protects the circuit under fault condition. It could be tested during the approval testing of the product and accepted for use on that product.	IEC 127/BS 4265
Isolators/ Interlocks	Provide similar operation as the mains switch.	IEC 947-3
Mains power switch	Most standards require the use of a manually operated mains switch as it provides protection against electric shock and fire — must be suitable for the application.	Product specific standard
Mains plugs	Provide protection against electric shock.	BS 1363/A
Power supplies	Provide regulated low voltage.	EN 60950 EN 61204
Mains power cords	Provide protection against electric shock — only approved types should be used. The ⟨HAR⟩ mark on the inner or outer sheath confirms the component is approved.	IEC 227/IEC 245 Other cables: EN 60719/799 HD 21 ... series HD 22 ... series HD 505 ... series HD 586 ... series
Opto isolators	Normally used in feedback or control circuits for power supply, bridging high voltage (HV) and low voltage (LV) circuitry.	Product specific standard
Plugs and sockets	Provide protection against electric shock.	Industrial: EN 60309-1/2 HD 196 S1
Printed wiring boards	Primarily provide protection against fire and electric shock — must be made of flame retardant material. Base material must be approved.	Product specific standard
Rating labels	Must be made of durable material and provide legible and indelible information (typically maker's name, Class II symbol (if applicable), supply voltage, frequency, nature of the supply (i.e. AC, DC), etc.)	Product specific standard
Relays	Could be used in AC mains switching operations or between AC and low voltage areas — in these cases an approved type should be used.	Product specific standard IEC 328/ BS 3955

Table 9.1 Examples of Safety Critical components (*cont.*)

Component	Comment	Possible standard
Switch mode power transformer	It bridges and provides isolation between hazardous (HV) and LV circuitry and provides protection against electric shock and fire. It must meet constructional and insulation requirements.	Product specific standard IEC 742 EN 61050 EN 60742
Temperature limiters/ thermal cut-outs	May be used in addition to fuses or in place of fuses; they are distinct from temperature control or thermostatic devices. Temperature limiters are not intended to operate in normal use.	Product specific standard, i.e. EN 60065/ BS 3995
User instructions	Must contain any information necessary for the safe use of the product (e.g. minimum ventilation distances, etc.) – must be in the official language of the market country.	Product specific standard
Other devices: Programmable controllers		EN 61131-2
Connecting devices	Various, e.g. clamping units, connection terminals, etc.	EN 60998-1 series EN 60999 EN 61210
Switches		EN 61058-1/- series EN 61095

9.5 Summary

Using approved Safety Critical components has the main advantage of reducing product test time and costs associated with the testing and approval of parts, product failure and retest, reworking, etc. It could also help to improve the reliability of the product. In addition, it would make the construction of the Technical File much easier and would help to demonstrate 'due diligence'.

The main disadvantage is the probable increase in the cost of the approved part compared to that of the non-approved type. For standard 'off the shelf' approved components, the cost difference could be minimal, but in cases where the vendor is requested to obtain component approval for a particular customer or application, the cost could be significantly higher.

Chapter 10

Designing for safety

An attempt is made in this chapter to further expand upon the basic safety principles mentioned earlier (in Chapter 8), and to assist the reader in gaining familiarity with the most common requirements quoted in safety standards. Furthermore, typical test and confirmation methods will be explained, thus making the process of self-assessment easier. However, it must be understood that the limits and test methods described in this chapter are 'typical' and (therefore) very general, and cannot be used alone for any one product's complete evaluation. Detailed test methods, limits and pass/fail criteria can only be found in the product specific standard, a summary of the test described in one of such standards is given in Appendix 6. Following the guidelines described here may therefore not be entirely sufficient to guarantee the complete and accurate product evaluation which may be necessary to ensure the supply to the user of a 'safe' product as required by the Low Voltage Directive.

Initially, the reader will need to understand that every safety standard specifies unique requirements for the testing of a particular product or product range, and for a given environment of use. It is clearly impossible to consider all of these requirements individually in this text; an attempt is therefore made to give the reader an understanding of those design and testing considerations which are applicable to *most* electrical products.

As mentioned already, testing according to a Harmonized standard is the best approach. However, since the interpretation and understanding of standards is an art in its own right (even experts often cannot agree on meanings and definitions), the testing has to be planned and carried out very carefully – especially if the manufacturer intends to test or evaluate against the Harmonized standard in-house, without external support.

A careful study of the requirements of the product specific standards, and due regard to the advice given in this chapter, will surely simplify the completion of a self-evaluation report.

10.1 Definitions (based on extracts from BS EN 60065: 1998)

Hazardous live – electrical condition of an object from which a hazardous *'touch current'* (electric shock) could be drawn.

Accessible part denotes a part that may be touched by the standard test finger (see Appendix 11).

Creepage distance denotes the shortest distance along the surface of an insulating material between two conductive parts (see Appendix 9).

Clearance denotes the shortest distance in air between two conductive parts (see Appendix 9).

Class I apparatus denotes an equipment in which protection against electric shock does not rely on basic insulation only, but which includes an additional safety precaution such that means are provided for the connection of accessible conductive parts to the protective (earthing) conductor in the fixed wiring installation so that accessible conductive parts cannot become live in the event of a failure of the basic insulation.

Class II apparatus denotes an equipment in which protection against electric shock does not rely on basic insulation only, but which includes additional safety precautions such as double insulation or reinforced insulation, there being no provision for protective earthing or reliance upon installation conditions.

Class III apparatus denotes an equipment in which protection against electric shock relies upon supply from an SELV circuit and in which hazardous voltages are not generated (extracts from BS EN 60950: 1992).

Basic insulation – insulation applied to a hazardous live part to provide basic protection against electric shock (Class I).

Supplementary insulation – independent insulation applied in addition to basic insulation in order to provide protection against electric shock in the event of failure of the basic insulation.

Double insulation – insulation comprising both basic insulation and supplementary insulation (i.e. is made up of two separate layers of insulation completely enclosing the live parts).

Reinforced insulation – single insulation applied to hazardous live parts which provides a degree of protection against electric shock equivalent to double insulation.

10.2 Equipment classification

During the design stages of a new electrical product, the designer has to take many important factors into account (design timescales, performance criteria, manufacturing costs, etc.). One of the most important considerations when designing to meet the safety criteria is the necessary level of protection against electric shock. As a very minimum the designer has to consider:

- the intended environment of use
- the operator's background/ability
- the operator's technical understanding
- the product's external construction.

This decision should be made very early in the concept stages as it could have significant design, testing, manufacturing and cost implications.

For equipment with metal enclosures, a protective earth wire is usually provided which is electrically connected to all exposed metal parts. In this case, if the primary (basic) insulation breaks down, the grounded enclosure will then trip the circuit breaker (i.e. blow the fuse).

If an ungrounded enclosure is used (e.g. a plastic casing), the standards recognize the inherent double insulation effect, i.e. the basic insulation between the primary supply and the enclosure as well as the additional 'supplementary' insulation provided by the enclosure itself. In this case, if the basic insulation breaks down, the enclosure will act as the secondary level of protection.

The classification of the product will determine the number of layers of insulation, and their types and thickness as well as the minimum distance that needs to be achieved between live and user-accessible parts. The type of testing which will be performed at a test laboratory will also depend on the type of insulation employed.

10.2.1 Class I

The safe operation of Class I equipment relies on the integrity of the external installation's earth system. On Class I products, live parts operating at hazardous voltages are protected by basic insulation, while accessible metal parts should be reliably connected to the safety earth.

Resistance between the main earth terminal and all earthed accessible parts must be typically less than $0.1\,\Omega$ – the resistance of the power cord is not included in this value. For some standards, the overall resistance between the earth pin of the mains plug and the EUT (Equipment Under Test) measurement point, must not exceed $0.2\,\Omega$. EN 60950 describes a measuring method by using a 12 V AC supply at 25 A or 1.5 times the current capacity.

To help the reader with the design of a good earthing, some of the most important considerations are listed below:

- The PE conductor shall not contain switches or fuses.
- The PE conductor can be bare or insulated.
- Disconnection of the PE at one assembly shall not break the grounding to other assemblies unless the hazardous voltages are removed.

Designing for safety

- In cases where the PE conductor is insulated, the coating should be green/yellow and should be secured to the chassis or frame via a closed-loop ring connector placed over a No. 6 (or larger) welded stud.
- For secure connection, it is advisable that a star washer be placed underneath and a lock washer on top of the ring connector.
- The PE should be secured by its own nut; other grounds can share the same stud but must be placed over the nut holding the earth ground conductor in place.
- The PE conductor must be mechanically secured before being connected to the ground pin of a coupler, e.g. an AC inlet. The same applies in cases of directly connected power cords.
- PE disconnection is not necessary unless a particular part is to be removed.
- The earthing connection conductor should be resistant to significant corrosion; the use of plastic coating might be necessary (EN 60950 gives a useful table of electrochemical potentials that exist between metals – Annex J, Table J).
- All operator-accessible metal parts must be electrically and reliably connected to the earth ground.
- A fuse or another overcurrent device should generally be provided in the primary circuit and should be connected in the hot (live) supply conductor.

Figure 10.1 Example of a good earthing connection

Testing for earth continuity must be performed as explained in Chapter 14. Equipment for this test varies in price and performance – the choice of equipment (and the purchase price) will depend on the test current requirement as specified in the product standard, and the perceived need for automated testing at the end of line.

10.2.2 Class II

It is not true to say that all equipment with an outer metal enclosure must be of Class I construction. If the metal enclosure is not grounded (i.e. Class II) but well protected by double or reinforced insulation, then the possibility of insulation failure (resulting in the enclosure becoming live) is negligible.

Accessible low voltage areas (typically less than 30 V AC or 60 V DC) in a Class II product which may be touched by the operator must be separated from hazardous voltage areas by full Class II insulation thickness, and creepage and clearance distances equivalent to denote insulation (values will vary between standards).

If an SELV (Safety Extra Low Voltage) circuit is used in a Class II product, voltages on accessible parts should not exceed typical values of 25 V AC or 60 V DC (lower voltages apply in special cases, usually depending on the environment of use).

While a Class I product relies on the integrity of its external earth system, Class II equipment is inherently safe by design. Class II equipment has become very common nowadays (e.g. domestic appliances, television receivers, video recorders, garden tools, electric drills, etc.), mainly because a plastic enclosure is lighter to carry and is more aesthetically pleasing.

10.2.3 Class III

Unlike Class I and Class II products where protective earth and insulation are required to prevent the user from the risk of electric shock, Class III products rely on the use of a SELV supply having an upper limit of 50 V AC or 60 V DC – reference EN 60950. This type of equipment is outside the scope of the LVD but many of the standards relevant to the LVD contain requirements for Class III products. Class III equipment is usually battery operated.

SELV circuits must be safe to touch under normal operating conditions and under single component failure. Accessible voltages between:

- any two parts of the SELV and
- any part of the LVD and earth

shall not exceed 42.4 V peak or 60 V DC under normal or fault conditions.

EN 60950 describes three methods for designing SELV circuits, these are given below:

- Method 1: Separation of SELV circuits from high voltage circuitry by double or reinforced insulation.
- Method 2: Separation of SELV circuits from other circuitry by earthed conductive screen or parts.
- Method 3: Adequate earthing of the SELV circuits, i.e. the use of a protective earth.

Not all product construction clearly falls under just one of these three classes – some may employ a combination of two construction classes while others may combine all three. An example of this would be handheld battery operated equipment (Class III construction) which is also provided with an add-on mains operated supply or charger (Class I or II). The insulation requirements for use in Class III mode are clearly different from those in the mains powered mode.

10.3 Insulation

All insulation used in an equipment must be fit for purpose – it has to be made of a material with a flammability classification and material deformation properties suitable for the place of use. For example, materials used for structural purposes or for the enclosure need to be of a much higher flammability rating than the insulation of live parts. Also, insulation must not be hygroscopic (material that absorbs water or moisture) as it will fail insulation testing following humidity treatment – see 10.10.

10.3.1 Basic insulation

This is used between live parts and earthed metal parts (Class I), and between live parts of different polarities – it is required by most standards for mains operated equipment. This type of insulation is not fail-safe and should not be accessible to the user. In cases of mains operated equipment (230 V–240 V) the insulation must withstand an electric strength test (i.e. there must be no breakdown of insulation). The test voltage may be 1350 V AC or up to 2500 V peak, depending on the applicable product standard.

10.3.2 Supplementary insulation

This is mainly used on live parts in Class II equipment to protect the user from electric shock. Supplementary insulation is a layer of insulation in addition to basic insulation, when it cannot be considered safe on its own. It is required to withstand an electric strength test and in the case of mains operated equipment, the test voltage should be 1500 V rms. The thickness of such insulation will vary depending on the applicable standard, for example the standard for domestic electrical appliances specifies a minimum thickness of 1 mm, while the standard for domestic electronic appliances, broadcast receivers, radios, etc. specifies a minimum thickness of only 0.4 mm.

In summary, supplementary insulation could be:

- two layers – each passing electric strength test of 1500 V rms
- three layers – any two must pass electric strength test of 1500 V rms.

10.3.3 Double insulation

As explained earlier this is an insulation comprising of basic and supplementary insulation, and is safe to touch (the mains power cord is a typical example of double insulation).

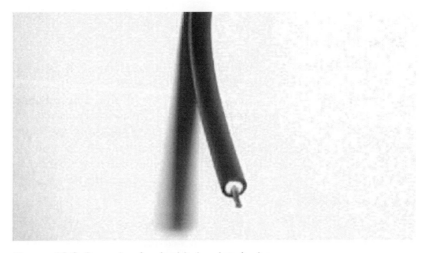

Figure 10.2 Example of a double insulated wire

10.3.4 Reinforced insulation

This is a single layer of insulation used to cover live parts. It can be found mostly in Class II products, and it is required to withstand an electric strength of up to 4000 V AC. As for supplementary insulation, the minimum thickness of reinforced insulation varies according to the applicable standard, it could vary from 0.4 mm (EN 60065) to 2 mm (EN 60335).

Protection of the user from the risks of electric shock is paramount and the product designer must give this very careful consideration. Electrical safety standards operate on the principle that if one level of insulation fails, then there is another means of protection which is unlikely to fail.

Table 10.1 Example – values of insulation thickness, recommended distances and test levels (extracts from BS EN 60065: 1994)

	Class I		Class II	
	Between 'live' and earthed parts	Between 'live' and unearthed parts	When double insulation is employed	When reinforced insulation is employed
Insulation distance	min. 3 mm	as Class II	between primary and secondary circuit min. 6 mm (3 mm + 3 mm)	between primary and secondary circuit 6 mm
Insulation thickness	min. 0.4 mm or pass dielectric strength test (>1.5 kV rms)	as Class II	0.4 mm + 0.4 mm or pass dielectric strength test (>1.5 kV rms)	2 mm or >0.4 mm if pass dielectric strength test (>3 kV rms)
Dielelctric strength test	min 1.5 kV rms	as Class II	min 3 kV rms	min 3 kV rms

This is known as the principle of 'double improbability'. A single level of insulation is therefore not sufficient to protect the user from any hazardous voltages. For a product to be safe, insulation must be provided by either:

- double insulation
- reinforced insulation or
- basic insulation + protective earth.

Confirmation testing for insulation *on wiring* can be performed relatively easily, using an AC high voltage source (possibly the same tester as used for end-of-line testing). The specified voltage is applied between the test cable's inner conductor and the outer part of its insulation (wrapped with metal foil or similar) for a typical duration of 1 min – insulation breakdown should not occur.

10.3.5 Bridging insulation

When the circuit design requires it, double or reinforced insulation separating primary from secondary circuits may only be bridged by approved components (approved for safety critical operation) where it is necessary to provide coupling or a control signal in switch mode power supply circuits.

Components likely to bridge such insulation include:

- resistors
- opto-couplers
- transformers
- capacitors.

For component safety approval standards, see Chapter 9.

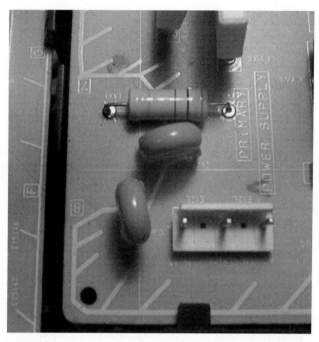

Figure 10.3 Example of a safety component bridging the HV/LV barrier

10.3.6 Creepage and clearance

Most safety standards stipulate a minimum spacing through air (clearance), and a minimum distance over surfaces (creepage) between primary (live) and secondary (low voltage) circuits. Creepage distances and clearances are dimensions which relate to possible breakdown paths between live parts and other components (see also Appendix 9). The clearance distance between two points is the shortest distance measured through air between:

- two conductive parts or
- a conductive part and bounding surface of the equipment.

To determine the equipment's minimum clearance requirements, the following have to be defined:

- the circuit, e.g. primary, secondary

- the pollution degree value (1, 2 or 3 where 2 is for normal working environments) – see Table 10.2
- the insulation working voltage
- the insulation category, e.g. basic
- whether or not the manufacturing process is subject to a quality control programme.

The creepage distance is the shortest path between:

- two conductive parts or
- a conductive part and bounding surface of the equipment measured along the surface of the insulation.

To determine the equipment's minimum creepage requirements, the following have to be defined:

- the circuit, e.g. primary, secondary
- the pollution degree value (1, 2 or 3 where 2 is for normal working environments) – see Table 10.2
- the insulation working voltage
- the material group (I, II or III)
- the insulation category, e.g. reinforced.

Creepage distances are not allowed to be less than the clearance.

Figure 10.4 Print side HV/LV circuitry separation

Not all live parts need be surrounded or separated by a layer of insulation. If the live part is fixed in place and is a sufficient distance away from other parts, then an air gap alone may suffice.

The creepage distances for basic and supplementary insulation will also depend on the material used in the Printed Wiring Board (PWB) and its

Table 10.2 Definitions of pollution degree

Pollution degree	Environment
1	Applicable for equipment sealed to exclude dust and moisture – only possible if a hermetically sealed enclosure is used.
2	Equipment subject to non-conducting deposits and some temporary condensation – this is the most common situation.
3	Equipment subject to conductive deposits (e.g. carbon brushes), pollution and condensation.
4	Equipment subject to conductive deposits of dust or rain.

'comparative tracking index' (CTI). As a general rule, most (PWB) materials of flammability rating 94V-0 (see Section 10.5) will be of Material Groups IIIa/IIIb – see below:

Material Group I: $600 \leq CTI$
Material Group II: $400 \leq CTI < 600$
Material Group IIIa: $175 \leq CTI < 400$
Material Group IIIb: $100 \leq CTI < 175$

Concentrating on the most common Pollution Degrees 2 and 3 and PWB Material Groups I, II and III, a table of minimum distances (in mm) given in EN 60950: 1992 is shown in Table 10.3.

Where secondary circuits are separated from primary circuits by an approved transformer, the minimum creepage and clearance distances in the secondary may, depending on the standard used, be less than those in the primary. This will depend on the security of the parts and the PWB

Table 10.3 Operational, basic and supplementary insulation – min. creepage (in mm)

Working voltage up to and including V rms or DC	Pollution degree 2			Pollution degree 3		
		Material			Material	
	I	II	IIIa & IIIb	XI	XII	IIIa & IIIb
50	0.6	0.9	1.2	1.5	1.7	1.9
100	0.7	1.0	1.4	1.8	2.0	2.2
125	0.8	1.1	1.5	1.9	2.1	2.4
150	0.8	1.1	1.6	2.0	2.2	2.5
200	1.0	1.4	2.0	2.5	2.8	3.2
250	1.3	1.8	2.5	3.2	3.6	4.0
300	1.6	2.2	3.2	4.0	4.5	5.0
400	2.0	2.8	4.0	5.0	5.6	6.3
600	3.2	4.5	6.3	8.0	9.6	10.0
1000	5.0	7.1	10.0	12.5	14.0	16.0

materials used, particularly since the CTI and flammability ratings will be critical.

Provided that due care is taken, the measurement of creepage distance and clearance is fairly simple. If the maximum advantage is to be taken of concessions in the standards then due care must be taken when following the rules of measurement. Careful handling of all equipment used for these measurements is vital, as is regular calibration.

Other factors affecting creepage distances and clearances are:

- the presence of moisture
- the materials used
- the construction and topography of the surfaces
- the consequences of any failure
- the susceptibility of the power source to surges in voltage.

Allowance must also be made for production tolerances – failure to meet the limits due to manufacturing tolerances, tool wear, shrinkage, etc. is not acceptable in law.

When setting limits for creepage and clearance, factors such as those described above are taken into consideration, but most requirements are based on the experience of the committees which set the standard rather than on fundamental laws.

10.4 Construction – wiring

Wiring used in electrical equipment can be categorized as internal or external. Internal wires are those used inside the equipment for PWB/sub-assembly connections, while external wires are those usually carrying hazardous voltages, i.e. mains power cord.

10.4.1 Internal wiring

Internal wires shall be fit for purpose. When deciding the type of wire to be used, the designer needs to consider its:

- rated voltage
- rated current
- operating temperature (relative to the parts it serves and may touch)
- support and possibly clamping/securing method.

10.4.2 Internal wiring – mechanical fixing

Internal wires have to be 'dressed' and secured so that the point of connection will not be subjected to excessive strain, loosening or damage. The construction must be such that if a wire becomes detached, the creepage distances and clearances are not reduced (by the natural movement of the wire) below those required in the standards (e.g.

6 mm between primary and secondary in the case of a Class II household appliance).

This requirement is met if there is no possibility of the wire becoming detached. This is arranged by wires having a mechanical as well as an electrical fixing – examples of a mechanical fix are wires twisted together, fastened together with tape, having a wrap joint, glued to PWB, etc.

Figure 10.5 Example – internal wiring dressing

Soldered wires must be mechanically secured before soldering – it must always be assumed that a solder joint by itself will fail. Wires connected to screw terminals should include a solder lug with upturned ends. For wires passed through metal guides, rounded smoothed edges or protective bushes should be used. Wires subject to movement should be protected from damage by some mechanical means, e.g. helical supports.

10.4.3 External wires and mains plugs

The mains (AC) power cord may either be connected to the product, or supplied in a detached form; in both cases an approved type must be used. A power cord is considered safe if it is approved by the international HAR or BASEC approval bodies – it must of course be suitable for the application for which it is intended and must have the correct number of conductors, i.e.:

- two conductors for connection to a Class II product (live and neutral)
- three conductors for connection to a Class I product (live, neutral and safety earth)
- three or more for 3 phase supplies.

Figure 10.6 Example of an approved mains lead

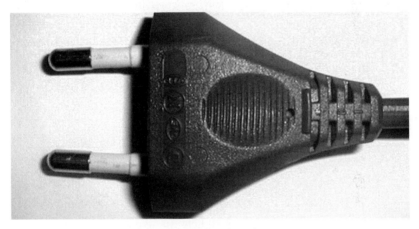

Figure 10.7 Example of an approved mains plug

The power cord's internal wires must also be insulated with the correct colour code as explained below:

- Live conductor – Brown
- Neutral conductor – Blue
- Earth conductor – Green/Yellow.

Note: *For permanently connected equipment, the colour code may be as above or as given below:*

- *Live conductor – Red*
- *Neutral conductor – Black*
- *Earth conductor – Green/Yellow*

The plug has to be rated to at least 125% of the rated current of the equipment, and some standards specify maximum length – this should be confirmed before deciding on the type to be used.

A fuse is connected in the 'live' side of the UK (3-pin) mains plug – attention should be paid to matching the rating of the fuse fitted to the plug and the fuse fitted in the product.

Mains (flexible) power cords for light use (e.g. household appliances) usually comply with IEC 227 (polyvinyl chloride insulated cables) or IEC 245. Conductors in power cords must be of adequate cross-sectional area such that if a short-circuit occurs in the equipment, the protection circuits in the electrical installation operate before the cord overheats.

EN 60335 specifies current ratings of mains cord conductors as given below.

Rated current of the equipment (amp)	Nominal cross-sectional area (mm^2)
<0.2	tinsel cord*
<0.2 and <3.0	0.50
<3.0 and <6.0	0.75
<6.0 and <10.0	1.00
<10.0 and <16.0	1.50
<16.0 and <25.0	2.50
<25.0 and <32.0	4.00
<32.0 and <40.0	6.00
<40.0 and <63.0	10.0

* Only to be used for cords up to 2 m in length (mains plug to equipment entry point).

Attention must be paid to the mains power cable entry to the equipment (smooth apertures will help to avoid cord damage, and bushings of insulating material must be of a durable type and should not deteriorate in normal use). An inspection check is advised, and an ageing test for bushings should be applied. The use of a cord guard is necessary in cases where the equipment is supplied with a flexible cord and is meant to be moved frequently. Tests for such cord guards will depend on the applicable standard (usually a flexing test at a certain angle and for several thousand movements).

Flexible power cords must be adequately anchored to the equipment, connecting points must be relieved from strain, and the conductors must be prevented from twisting. Cord clamps should be carefully designed so as:

- to be suitable for the power cord used
- not to damage the cord (sharp edges, too much compression, etc.)
- to withstand a pulling test (see below)
- to be made of insulating material (for Class II equipment)
- to be provided with lining (if they are of metal construction).

10.4.4 Internal wiring – insulation

Internal wiring should:

- be of adequate cross-sectional area
- be of adequate flame resistance
- be protected by a layer of insulation depending on its application.

Primary or high voltage secondary wires likely to be touched by the operator or the service engineer must have a secondary level of protection, e.g. a second layer of insulation such as tubing; otherwise, the whole unit must be provided with some other form of protection such as an interlock.

The insulation needs to withstand a dielectric strength test that will depend on the wire's application and the applicable standard. It may need to withstand only 1500 kV rms (measured between the conductor and foil wrapped on the outside of the cable).

Internal wire insulation temperature rise is normally limited to 50°C, but some internal wires may need to operate at much higher temperatures, e.g. those placed near high temperature operating parts or components – in such cases consideration must be given to:

- a reliable mechanical fixing (keep as far away as possible from the source of heat)
- the insulation material and its properties (use of low flammability grade, e.g. V2 or better)
- the possible use of additional sleeving (again with low flammability rating)
- the use of insulation which has a high temperature rating (ensure that rating is sufficient for the expected temperature rises).

Some standards require that internal wires be insulated with PVC, PFE, PTFE, FEP or neoprene, or marked with a rating of VW-1.

In summary, internal wires must:

- be provided with an adequate electrical insulation
- have adequate cross-sectional area (depending on the current they will carry)
- be of the correct voltage rating
- be of an adequate temperature rating (depending on its application)
- be of a correct flammability rating (depending on the applicable standard)
- be reliably fixed – shielded from hot parts, sharp edges, etc.
- be provided by additional sleeving (if necessary) for additional electrical, temperature or flammability protection
- be provided with sufficient strain relief to avoid damage and potential short-circuit
- to be provided with lining (if they are of metal construction).

Equipment fitted with a non-detachable power cord must be fitted with a suitable means of securing the power cord, e.g. a strain relief grommet.

The power cord should not be capable of being pushed back into the equipment to the extent that the cord or its conductors be damaged, or internal parts or components become loose, displaced or damaged. It should therefore be securely fixed to the equipment; this can be confirmed by performing the following simple pulling test (see product standard for details).

A pulling force of 40 N is applied to the mains cord 100 times and for 1 second duration. During this test the power cord should not become damaged, and a maximum of 2 mm displacement of the power cord is allowed. This is checked by visual inspection and by confirming that the internal creepage and clearance distances are maintained.

Although some standards specify a pull test independent of the equipment's weight, others specify a pulling force which is weight dependent:

- 30 N (newton) for equipment of weight less than 1 kg
- 60 N (newton) for equipment of weight between 1 kg and 4 kg
- 100 N (newton) for equipment of weight greater than 4 kg.

In addition to the pull test, the mains cord is subjected to an applied torsion of 0.25 Nm for 1 min – there should no visible damage or ripple.

The values given above will vary depending on the applicable standard and the mains cord size.

10.4.5 Withdrawal of mains plug

There must be no risk of shock when touching the pins of the mains plug after it is withdrawn from the socket outlet. A relatively simple and typical method to confirm this point is explained below.

With the mains switch in the most unfavourable position, the pins of the plug must be measured not to be live 2 seconds after the plug has been withdrawn. The test should be repeated up to 10 times to cover the most unfavourable condition.

10.5 External construction

Product safety standards contain construction and performance criteria. Under normal operating conditions the construction of the product should be such that the user will not be faced with the risk of electric shock.

Depending on the standard to be followed and the product's environment of use, there are a number of different tests that can be used to assess accessibility. The object of all such tests is to ensure that live parts do not become accessible.

Test methods specified by some of the most 'popular' standards are described below:

EN 61010 – specifies that the Rigid Test Finger (see Appendix 11) be applied to all surfaces including the base (some exceptions are listed) with a force of 10 N. There is no need for the application of the Tapered Test Pin (see Appendix 11) but the standard specifies the use of a 100 × 4 mm diameter pin for assessing vents and a 100 × 3 mm diameter probe for assessing controls.

EN 60950 – specifies that the Rigid Test Finger be applied to all surfaces of the equipment with a force of 30 N; floor standing equipment of weight >40 kg should not be tilted for this test (some exceptions are listed). The Tapered test is applied to all surfaces (without removing operator removable parts).

EN 60335 – specifies that the Rigid Test Finger be applied to all surfaces of the equipment with a force of 30 N; floor standing equipment of weight >40 kg should not be tilted for this test (some exceptions are listed). For Class II products, the Test pin is used – but not to lamp sockets with lamps removed and not to socket contacts. This standard also requires the use of some special tools such as a 30 mm diameter conical test pin.

EN 60065 – the requirements of this standard are more onerous as these products are usually of Class II construction, with plastic enclosure and used in a domestic environment where children may be playing. Some of these tests are described below (see also Appendix 6).

The standard specifies that the Rigid Test Finger be applied to all surfaces including the bottom with a force of 50 N and the Tapered test pin be applied to all outer surfaces. Other tests include the use of a 2 mm diameter chain for live shafts, a 100 × 4 mm diameter probe for ventilation slots, a 100 × 2 mm probe for preset controls and a 100 × 1 mm wire for bushings.

The enclosure must also be resistant to external forces. It provides the first line of defence against shock, fire and other hazards. Barriers are placed inside the enclosure which prevent the operator from reaching hazardous areas of the circuit, while guards are extra pieces mounted on the outside of the enclosure – usually to protect the operator from mechanical hazards such as fans or gears.

Plastic enclosures should be tested for strength. One possible test would be to use the Rigid Test Finger for applying inward force and a Test hook for applying outward force (these tools are shown in Appendix 11). The object of the exercise is to assess whether or not distances between accessible parts and live parts are reduced. Creepage and clearance distances should be maintained and live parts must not become accessible.

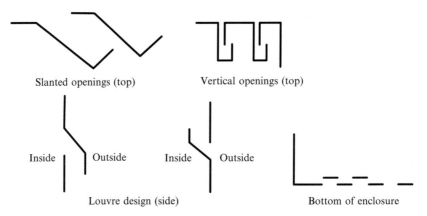

Figure 10.8 Examples of enclosure openings (extracts from BS EN 60950: 1992)

Note: *Care must be taken during the design of the enclosure to ensure that any openings which might be necessary for ventilation or access be kept to a minimum.*

Typical dimensions of openings for ventilation are 2 mm × 15 mm on the side of the enclosure, and 3 mm × 30 mm on the bottom of the enclosure. Openings for the power cord, preset controls, etc. should also be kept to a minimum. Ventilation openings directly above the power supply and other high voltage areas are not encouraged (to avoid hazard due to accidental liquid spillage).

In cases where ventilation openings above such areas are absolutely necessary, then vertical or slanted openings as shown in Figure 10.8 should be used.

10.5.1 Mechanical strength

Equipment must be constructed to withstand the handling expected in normal use. Most standards specify test methods for assessing conformity to mechanical strength requirements; some of these include:

Bump test – the equipment under test is placed on a horizontal table, and raised and dropped from a set height (e.g. 5 cm) a specified number of times in order to confirm the security of internal components. At the end of the test, a visual inspection is carried out – there should be no reduction in creepage and clearances.

Vibration test – this is applicable to portable equipment. The equipment is vibrated at a set amplitude for a short period of time. At the end of the test, the equipment must not have become unsafe due to reduction in creepage and clearance distances or the loosening of fixing screws or any other components.

Impact test – using the impact test hammer as shown in Appendix 2, the equipment is subjected to a number of blows to every point providing protection to live parts.

The impact is also applied to windows, lenses, etc. There must be no damage affecting the safety of the product, i.e. the enclosure must not have any visible cracks, live parts must not have become accessible and insulating barriers must not have been damaged.

10.5.2 Mechanical stability

Equipment must be stable in use and not liable to topple over and pose risk to the operator when in use (i.e. user being crushed underneath).

A simple confirmation method is to place the equipment on a plane inclined at 10 degrees to the horizontal and rotate the product through 360 degrees. During the rotation, a force of 100 N directed vertically downward at any point, on any (normally) horizontal surface (to provide the maximum overturning moment), should not cause the equipment to fall over.

For equipment more than 1 m in height, and weighing more than 25 kg, a force equivalent to 1/5 of its weight (max. 250 N) should be horizontally applied up to a height of 2 m on every side; once again the equipment should not tip over. The test should be repeated with doors, drawers and other parts in the most unfavourable position.

Other tests are described in the product standard.

10.6 Resistance to fire

Three ingredients are needed to support fire, i.e. fuel, heat and oxygen. Therefore, in order to prevent a fire from propagating, thereby becoming a hazard, most safety standards set maximum limits on fuel content, heat and the enclosure.

Enclosure, guards and barriers must not only be resistant to external forces as described above, but also must be made of materials that are not highly combustible – such as steel, aluminium, or heat resistant tempered, wired or laminated glass. When plastics are used as enclosures, barriers or guards, they must comply with the relevant flammability ratings – a typical rating for an enclosure of a large equipment (more than 18 kg), would be 94 5V-A (94V-1 would be acceptable for smaller equipment weighing less than 18 kg).

The flammability rating of polymeric materials is mostly tested against known standards such as the most commonly used Underwriters' Laboratories (UL) UL94 classification as explained below. The main tests for this classification fall into three categories:

(i) UL94HB (least stringent)
(ii) UL94V-0, UL94V-1, UL94V-2
(iii) UL94 5V-A, UL95 5V-A (most stringent).

All safety standards have different flammability requirements, this will largely depend on the product's environment of use, for example standards applicable to office and business equipment (EN 60950) have particular requirements for the flammability rating of all fire risk materials as described below.

Enclosures and large parts of the enclosure having ventilation holes designed for letting out heated air must be of slow burning or fire retardant material; they should typically be made with UL94-V0 (or better) material.

Printed Wire Board (PWB) material on which components are mounted should be of flammability UL94-V1 or better, and should be separated from less fire resistant material by at least 13 mm of air. PWBs with a surface area of less than 25 cm^2 are normally exempt from this requirement.

Air filter assemblies should be constructed of materials of flammability rating UL94-V2, or HF or better.

Decorative parts (parts of the enclosure whether mechanical or electrical) should be of flammability rating HB or better.

Small external decorative parts such as mounting feet and knobs may be exempt from the requirements as they are likely to have little or no effect on the spread of a fire.

Components such as integrated circuits, opto-couplers, capacitors, etc. are also exempt if they are mounted on V-1 or better material.

Internal plastics should have flammability rating UL94V-2, or HF-2 or better.

Internal wires should have flammability rating VW-1 or better.

High voltage transformers and components (above 4 kV) must be resistant to fire.

10.7 Electrical connections and mechanical fixings

Screw fixings which may be loosened and tightened during the life of the equipment must have adequate strength. This is usually checked by loosening and tightening the fixing screws 10 times to the specified torque (typically 1.2 Nm for enclosure screws). There must be no deterioration that could affect the safety of the product.

There must be means of ensuring correct location (by a guide in the screw hole) to prevent the screw angling away from its intended fitting direction. If the fixing screws are not captive then creepage and clearance distances are checked by using a screw with a length equal to 10 times the nominal diameter of the correct screw (typically 40 mm for a 4 mm diameter screw) at the torque specified above. There must be no reduction in creepage distances and clearances below those allowed.

Stands or detachable legs supplied by the manufacturer must be supplied complete with the necessary fixing screws to prevent the possibility of the customer using fixing screws that are too long – thus reducing creepage distances and clearances.

10.8 Components

Components which bridge the insulation between primary and secondary circuits and provide protection against electrical shock and other hazards should be of an approved type. These components are covered in detail in Chapter 9 but some additional information is also given below.

Fuse links must comply with a relevant standard, must have adequate rupturing capacity and be marked with their characteristics either on the fuse holder, or adjacent to it.

Figure 10.9 Example of fuse marking on PWB

Mains power switches must have an adequate contact gap and be of adequate rating. The equipment or the rack should be marked with O and 1 to identify the off/on positions.

Interlocks must be used if access to the inside of the equipment by the operator is necessary, e.g. to adjust moving parts, remove jammed paper, replace bulbs, etc., as during such operations, the possibility of the equipment becoming energized must be prevented. Such interlocks must open all primary current-carrying conductors that provide power to the accessible areas, and be self-restoring. Interlocks must be good for many thousands of operations and must be approved.

10.9 Temperature

Excessive rise in temperature can cause fires and/or damage the insulation used to isolate hazardous parts. Parts that typically exhibit temperature rise, such as transformers, relays, inductors, large electrolytic capacitors, switches, fuses (as well as the PWBs on which they are mounted), must be measured for temperature rise. Test method examples are given below.

10.9.1 Heating under normal operating conditions

Most standards describe a suitable test method that is intended to ensure that the equipment does not fail or become unsafe in use due to overheating. A typical test method for temperature confirmation is explained as follows.

The equipment is placed in the normal position of use and ventilation, in accordance with the instructions supplied by the manufacturer. If such conditions are not specified, then it should be placed in a cabinet (e.g. a wooden box) with a small gap between the sides of the cabinet and the equipment under test (possibly 5 cm all around). The equipment under test should (in order to determine the worst case) be connected in turn to supply voltages of 0.9 and 1.06 times the rated voltage as well as the nominal rated voltage itself. In normal operating conditions, and when a steady state has been reached (usually after 2 to 4 hours), the temperature rises are measured.

Temperature rises above ambient must not exceed a typical value of 85°C for PWBs and winding wires, and 60°C for the enclosure. The best method to measure temperature on components and PWBs, etc. is to use thermocouples – these are attached to those parts of the equipment which typically operate at the highest temperature such as:

- the upper part of enclosures (usually above the power supply)
- behind front cover flaps
- between a switch mode transformer winding and core
- deflection coils (if employed)
- voltage regulators
- line and frame output devices
- high current diodes
- audio output circuits, etc.
- insulation.

The enclosure is removed and the thermocouples are attached to a digital thermometer via a switch box, and the enclosure is refitted. For component testing, thermocouples should be soldered to their respective solder 'lands' on the PWB.

Elsewhere, high temperature adhesive tape may be used to attach thermocouples to the cabinet, or between transformer winding and core, etc. – in each case ensuring that the tip of the thermocouple is touching the point to be measured.

Note: *A preliminary operational run and examination of the equipment may need to be made in order to determine those parts which are most likely to reach high temperatures.*

10.9.2 Heating under fault conditions

All safety standards require that equipment be safe under abnormal operating conditions. Electronic circuits are tested by simulating component failure that might occur during normal use, thereby causing the equipment to become unsafe. In testing, only one simulated fault or abnormal condition is applied at a time.

Protection against electric shock must exist under fault conditions. The temperature of parts acting as a support or barrier must not reach unsafe levels (levels likely to cause mechanical failure), such that live parts become accessible or creepage distances and clearances be significantly reduced.

Furthermore, parts must not reach a temperature such that there is a danger from fire or abnormal heat, or that flammable gases are emitted. The latter condition generally applies only to PWBs for which the typical limit is a 110°C rise above ambient. This temperature limit may be increased provided that an area of not more than $2\,cm^2$ of the PWB becomes heated and solder does not become molten within that area.

As already indicated, the tests are carried out by simulating component failure and ensuring that under these conditions, parts of the equipment do not exceed permitted temperature rises and do not emit flammable gases.

From the test engineer's general experience, knowledge of the product type and study of the circuit diagram, simulated fault conditions are decided and applied to the product. Measurements are then taken of the temperature rise of relevant parts by the attachment of thermocouples. Thermocouples should be attached as described in 9.9.1 above.

Typical examples of fault conditions to be applied are as follows:

- S/C (short circuit) of capacitors connected between supply rails and ground
- O/C (open circuit) one of all current sharing pairs of high wattage resistors and/or coils connected in parallel
- S/C audio output connections

Table 10.4 Some typical maximum permissible temperature rises

Part	Permissible temperature rise (°C)	
	Normal operating conditions	Fault conditions
Accessible parts:		
Knobs, handles, etc.		
• metallic	30	65
• non-metallic	50	65
• metallic enclosures	40	65
• non-metallic enclosures	60	65
Electrical insulators		
• supply cords and wiring not under mechanical stress	60	100
Printed circuit boards (depending on type used)	85 ~ 120	110 ~ 150
Mouldings (depending on type used)	95 ~ 110	130 ~ 150
Support parts or mechanical barriers (wood and wood-based material)	60	90
Transformer winding wires (depending on type of insulation)	55 ~ 145	75 ~ 180

- S/C deflection coils (if employed)
- S/C across any two junctions of high power semiconductors
- S/C across parts of bridge rectifiers
- S/C the output of power supply units
- S/C output connections of transformers.

Following each simulated test, the equipment must remain safe within the context of the standard; i.e. there must be no increased risk of electric shock, fire or mechanical hazard. In addition to overheating due to electrical component failure, other common abnormal (simulated) tests should include:

- blocked vents
- stalled fan motors
- paper jam (copiers).

Protection against excessive temperature rise and overheating in fault (or simulated fault) condition is usually provided by a total interruption of the supplied power to the equipment by employing devices such as:

- fusible links
- thermal cut-outs

- fuses
- circuit breakers.

Some typical maximum permissible temperature rises (above ambient) under normal operating and fault conditions are given in Table 10.4, taken from BS EN 60065: 1994.

Built-in components, plugs, sockets and internal wires will have individual operating temperature requirements; these will either be marked on the component or indicated in the manufacturer's instructions – care must be taken in the design and assembly of the product so that they are not exceeded.

The values of permissible temperature rise will depend greatly on the product, the applicable standard, the environment of use (e.g. in high or low humidity location), the class of the insulating material used, the construction of the product and other factors.

Each standard will quote different values and it is up to the manufacturer to confirm that these values are not exceeded – not only during normal operation, but also throughout the assembly process (e.g. soldering), and under abnormal (fault) conditions. Only then can a decision be made as to whether the product is safe for release to the market.

10.10 Dielectric strength

Dielectric withstand tests are applied to all products in order to evaluate the efficiency of their electrical insulation. Also, the equipment under test must not become unsafe due to the humidity conditions which may occur in normal use and cause hygroscopic insulation to break down.

A typical test method would be to place the equipment in an environmental chamber for 48 hours at a typical temperature of 30°C, and 90–95% relative humidity (RH). This would be increased to 40°C for up to 7 days (same RH) for equipment that will be used in tropical conditions. Immediately upon completion of the humidity treatment, a (mains frequency) AC test voltage of a substantially sine wave form, or a DC test voltage or a combination of both with peak value as specified in the standard (e.g. 3 kV rms for a Class II product) is applied for 1 minute across:

- the mains poles of the power supply – usually by connecting the poles at the mains plug (joined together) and
- accessible metal parts and/or input or output terminals (e.g. for the connection of peripheral equipment).

During the test, mains and functional switches should be placed in the 'on' position. No flashover or breakdown should occur.

Not all standards require preconditioning of the product before applying this test. Test voltage levels will vary for each product

Table 10.5 Dielectric strength test voltages (based on BS EN 60065: 1994)

Point of application of test voltage	Test voltage (V rms)		
	Class I	Class II	Class III
Between live parts and the body (if metal)	1250	3750	500
Between live parts and other inaccessible metal parts	–	1250	–
Between inaccessible metal parts and the body (if metal)	–	2500	–

category – some typical values applicable for most products are given in Table 10.5. In addition to initial approval testing, this test should be applied 100% to all finished products before dispatch.

The type test, which overstrains the insulation, is performed for a duration of 60 seconds, while the production test is performed for a minimum of 1 second.

10.11 Insulation resistance

The insulation resistance is measured at no less than 250 V AC or DC. Typically, a voltage of 500 V DC is applied. Some typical test points and insulation resistance values are given in Table 10.6.

In practice, the insulation resistance measurements are expected to be much higher than the values quoted above and accordingly the manufacturer's own rejection criteria should reflect this.

Table 10.6 Typical insulation resistance values

Insulation to be tested	Minimum insulation resistance (MΩ)
Between mains poles of the supply	2
Between mains poles of the supply and accessible parts and terminals	4
Between live parts and earthed accessible metal on Class I equipment	2
Between live parts and functionally insulated metal parts on Class II equipment	2
Between functionally insulated metal parts on the body of Class II equipment	5
Between live parts and metal parts on the body of Class II equipment	7

10.12 Earth leakage current

Excessive current flow may be hazardous to an operator. Under fault conditions (e.g. loss of protective earth, reversed polarity of the AC mains), the chassis may assume hazardous potentials and excessive leakage currents will inevitably occur. The test for earth leakage current is applied as per the method described in the applicable standard, the test voltage usually is:

- 1.06 times the rated voltage or 1.06 times the upper limit of the rated voltage range – if the rated voltage or upper limit of the rated range does not exceed 250 V. This applies to equipment operating (within this limit) on DC only, single-phase DC only and for equipment which has the capability of running on both single and three phase supply.
- 1.06 times the rated voltage or 1.06 times the upper limit of the rated voltage range, divided by 1.732.

The leakage current is measured within 5 seconds after the application of the test voltage.

Some typical values of maximum leakage currents are given below:

All Class II equipment	0.25 mA
Portable Class I equipment	0.75 mA
Stationary Class I equipment with heating elements which are detachable or can be switched off separately	0.75 mA/kW (5 mA max.)
Stationary Class I equipment	0.75 mA/kW (5 mA max.)
Movable Class I equipment	3.5 mA
All Class II equipment	0.25 mA

10.13 Measurement and measurement equipment

Most of the tests described so far in this chapter can be carried out in-house and without major expense to the manufacturer. However, for a manufacturer to self-declare compliance with a product specific standard, an in-depth knowledge of the requirements, the adequate competence of testing engineers, and the use of accurate and calibrated instruments will be essential.

When measuring parameters such as temperature and distance, calibrated instruments with known measurement uncertainties must be used. Although the uncertainties may seem small, they can possibly make the difference between compliance and non-compliance. For example, a temperature rise measured at 58°C or more using a system with a measurement uncertainty of 3°C, should be declared to fail to comply with a requirement for a maximum permissible temperature rise of 60°C.

When designing products, it is advisable to allow comfortable margins in distances (creepage and clearance), and maximum permissible temperatures.

When applying fixes for safety, the designer must not overlook the following:

- the customer's expectations of the product
- the electromagnetic compatibility requirements (and how design issues impact upon compliance)
- the reliability of components (many components, equipment and controls are required to undergo an endurance test).

By using high accuracy, regularly calibrated (traceable to National standards and covering the range of use) measuring instruments, a manufacturer can have confidence in the results of confirmation tests. Some of the most commonly used measurement instruments necessary for testing to confirm compliance to basic safety principles (as quoted by most standards) are listed in Appendix 5.

10.14 Protection of service personnel

Special consideration is necessary where equipment is of such size and complexity that it may be necessary for service personnel to reach over, under or around uninsulated electrical or moving parts. In such cases, all parts providing risk of entrapment or high energy electrical shock should be placed so that contact or bridging is unlikely. Components such as capacitors placed in the mains circuit must either discharge to less than 50 volts in a very short time (typically within 1 second), or be marked with a warning label. Similarly, parts that would normally be expected to be at ground potential but actually are not, must also bear warning labels.

10.15 Other hazards

10.15.1 Spillage

If in normal use liquid is likely to be spilt into the equipment, the equipment should be designed so that no hazard will occur, e.g. as a result of the wetting of the insulation or of internal uninsulated parts which are live.

10.15.2 Overflow

Liquid overflowing from containers in the equipment (due to overfilling) should not cause a hazard during normal use – e.g. as a result of the wetting of the insulation or of uninsulated live parts. Similarly, equipment that is likely to be moved while a container is full of liquid should be protected against liquid surging out from the container.

10.15.3 Liquid leakage

Equipment should be designed so that liquids leaking from containers, hoses, couplings, seals, etc., do not cause a hazard – e.g. as a result of the wetting of the insulation or of uninsulated live parts.

10.15.4 X-rays and cathode-ray tubes (CRTs) for TV and computer monitors

CRTs imply hazards from X-radiation and implosion. They must be of an approved type and tested in dedicated facilities. Twelve samples are tested in total (half after having undergone an ageing process). CRTs are tested for:

- *Mechanical strength* – impact from a hardened steel ball dropped from a height.
- *Implosion* – an area is scratched with a diamond stylus then cooled with liquid nitrogen until a fracture occurs.
- *Ionizing radiation* – the equipment's controls are adjusted to give maximum radiation while maintaining an intelligible picture. Radiation measurements are made after 1 hour.

10.15.5 Ultraviolet radiation

Equipment containing a UV light source, but which is not designed to provide external UV illumination, should not permit the unintentional escape of UV radiation that would be harmful to the operator.

10.15.6 Microwave radiation

Unintentional microwave radiation should not exceed safe limits (as given in the relevant product standards) in the vicinity of the equipment.
  In addition to the above, consideration must also be given to the effects of:

- sound pressure
- ultrasound pressure
- laser sources
- liberation of poisonous gases
- explosion and implosion.

10.16 Labels

Most standards require that where danger exists, warning and caution labels be fitted to the product. Words such as DANGER, WARNING and CAUTION are to be of sufficient height and should not be placed on parts likely to be discarded.

86   Electrical Product Safety

External labels should be placed where they can be easily seen when the equipment is in its normal operating condition; labels inside doors, covers, etc. are allowed if easily opened.

Voltage and current ratings of every fuse must be marked adjacent to the fuse, and a caution label added advising that only the same type of fuse should be used as a replacement. It is advisable that the exact marking as indicated on the fuse is also placed on the caution label. If an IEC 127 fuse is used, the fuse will be marked as per the example shown below:

T 5A L 250 V

The first letter or colour band indicates the fuse characteristics:

T: Time-lag or Blue
TT: Long time-lag or Grey
M: Medium time-lag or Yellow
F: Quick acting or Red
FF: Very quick acting or Black

5A: current rating

L: Indicates low breaking capacity (usually glass fuse body)
H: indicate high breaking capacity (usually ceramic fuse body)

In addition to the fuse caution label, warnings are also required where:

- hot surfaces may come in contact with the operator or with service personnel
- direct plug-in units are required but not supplied
- capacitors which may hold charge could come in contact with service personnel – the warning should advise how they are to be discharged
- X-ray or other hazards (e.g. laser) may exist
- chemicals are used
- there are residual mechanical hazards.

In addition to the above (warnings to the operator), it is reasonable to warn service personnel:

- to isolate the power before removing covers if live parts will be exposed
- which parts of the equipment contain hazardous voltages.

Service personnel will make certain assumptions when maintenance or repair work is carried out. These will include:

- unmarked metalwork and components are safe to handle
- fans hidden within the equipment will be fitted with finger guards
- any lifting operation will require only one person
- mechanical parts will not move and thereby cause risk of injury
- laser or other radiation is not present.

If any of the above conditions are not true, additional warnings will be necessary.

Note: *It is very important that safety measures be* designed *into the product. Reliance on operator warnings should be avoided wherever possible by (instead) ensuring that service operations (particularly when within the capability of the operator) are* inherently *safe. Safety interlocks which automatically interrupt power supplies, for example, are clearly preferable to labels advising the presence of high voltages if covers are removed or doors opened. Where warning labels* are *essential, they must of course be clear, and written in the appropriate language for the country or countries to which the product will be supplied.*

10.17 Markings

All markings and labels used should be indelible, legible, and placed on the exterior of the unit (excluding the bottom). The type identification marking (nameplate) should contain the following information:

- Manufacturer's (or responsible vendor's) name or trademark.
- Model number (or model name) and serial number.
- Symbol of the nature of supply, e.g. $\sim$ for AC power supply (1).
- Rated operating voltage or voltage range (2).
- Rated maximum power in watts (W) or volt-amperes (VA) or rated current (A or mA).
- Rated frequency or range of frequencies (3).
- Warning of the presence of hazardous voltage inside and advice to the operator that only qualified personnel should access the inside of the equipment.
- Safety agency approval marks, e.g. (if applicable) the CE Marking.
- Warning of possible presence of high leakage currents (for fixed equipment).
- The double square symbol for Class II equipment, i.e. ▣.

1. Voltage rating may be indicated as:

- $\sim$ or AC for alternating current supply
- ▬ or DC for direct current supply
- $3\sim$ for three-phase AC supply
- no marking is necessary if the equipment operates with *either* AC or DC supply (or $\tilde{-}$ might be used).

2. The rated operating voltage can be indicated as per the examples below:

- 110 V–240 V, if there is no need for voltage adjustment
- 115 V/230 V, if adjustment is necessary.

3. The equipment's operating frequency range can be marked as:

- 50 Hz–60 Hz, if there is no need for adjustment
- 50 Hz/60 Hz, if frequency adjustment is necessary.

Safety warning indications on the model name label should always be given in the language of the country of use.

A simple test for ensuring the durability of the rating label is to rub the label lightly for 15 seconds with a cloth soaked in water, then (in a different place) with a cloth soaked in petroleum spirit – the information on the label should remain legible in both locations, without the edges or corners curling up.

10.18 User instructions

Most equipment and products require instructions for use. All Class I and Class II equipment must be accompanied by user instructions detailing the method or steps necessary for connection to the supply. For Class I equipment an additional warning is necessary, i.e.:

> WARNING – This appliance must be earthed

Information necessary for the safe operation of the product could include:

- advice on the need for an RCD in connecting the equipment to the mains supply
- warnings regarding hazardous moving parts
- warnings regarding the presence of ultraviolet or infrared lamps
- installation instructions such as clearances for equipment to be fixed to the wall or onto the floor, etc.

User instructions must be simple to follow, and be written in the native language of the user. Attention must be paid to translation if the equipment is to be marketed in many countries – a phrase in one language may mean something else in another!

It is clearly desirable, therefore, that the translation work be done by a native of each country in which the product will be sold. The preparation and supply of user instructions is an invariable requirement of almost all safety standards, and their content and accuracy will be scrutinized as part of the evaluation process when products are submitted for testing to a third party.

Instructions for use should also include a repeat of all warning labels placed on the equipment with an explanation of how to minimize the risk.

Chapter 11

Compiling an Evaluation Report

Where possible, testing by an accredited laboratory to a Harmonized standard is a step every manufacturer should take before compiling his Technical File and affixing the CE Mark to his product. However, this approach is not an option for most small manufacturers; this is particularly true for those manufacturers which produce a large range of products in small quantities, or for those who build 'one-off' units according to a customer's specifications.

For those manufacturers unable to test their products by an accredited external laboratory, it is strongly suggested that internal product evaluation is performed as is was explained earlier in Chapter 7 and that an Evaluation Report is compiled and retained as a part of the Technical File.

The objective of compiling an Evaluation Report is to prove to the manufacturer himself and to the appropriate enforcement officer that the manufacturer has:

- taken all reasonable steps
- followed good engineering practice
- made a comprehensive evaluation of the product
- designed a safe product
- minimized the risk of failure while in use.

11.1 Product specific reports

Over many years, CENELEC in co-operation with the National Certification Bodies has created standard report formats for the evaluation of products for a number of safety standards. Such report formats would be used by all NCBs to ensure the repeatability of tests and evaluations suggested in the Harmonized standard and would ensure their acceptance by other CCA/CB scheme members.

Today, standard Evaluation Reports exist for some product specific safety standards as listed below:

EN 61010 – Laboratory equipment
EN 60950 – Information technology equipment
EN 60065 – Broadcast equipment
EN 60335 – Household equipment

Manufacturers of products within the scope of these standards and wishing to self-evaluate, are quite fortunate as they can use the standard report format (obtain a copy of the local NCB) and follow it step by step. It may not be possible to perform all the prescribed tests 'in-house', as this will depend on the technical expertise and equipment availability.

11.2 Own design reports – contents

Manufacturers of products other than those covered by the above standards have a more difficult task in deciding:

- what to test
- how to test
- how to record their evaluation data.

For manufacturers of products not covered by Harmonized standards, we believe that they can either:

- create and follow their own report format (based on what they feel is necessary to test or evaluate to meet the demands of the LVD) or
- follow a particular Harmonized standard, design a report format suitable to the manufacturer by listing the requirements of the standards and follow it step by step during product evaluation.

To help the reader design a report, we have prepared an example report as given below and we have listed evaluations for some of the main safety principles quoted in most safety standards without going into detailed tests and evaluations. The example report below follows a layout similar that of a standard CENELEC report but is not based on any one particular standard.

If the reader now wishes to use this format as the basis for his product evaluation, he:

- must not forget that the test and evaluations suggested are only examples
- should modify this report to include his product's individual product standard's or regulation's requirements.

It has to be noted here though that the Department of Trade and Industry (DTI) in the UK recommends that in cases where

Harmonized standards are not applied, the manufacturer should seek assistance from a Notified Body or another qualified agency.

By preparing the example in 11.3, the authors' aim is to give the reader an 'idea' only of how to generate a report format. If the example is to be followed, it has to be extensively modified to include the requirements of the product specific standard or regulation applicable to the manufacturer's product. If this is not possible, then it must list any special precautions or measures that has been taken and the evaluations performed to ensure the product's safe operation. The report format suggested as it stands does not provide a 'generic solution' for any product or product category, to create such a document we believe would be 'dangerous' to those inexperienced with testing for electrical safety.

11.3 Product Evaluation Report – example

Report No: 001   Page 1 of 21

PRODUCT EVALUATION REPORT	
Based on Safety Standard: *e.g. BS EN 61010-1:1993 – 'Safety requirements for electrical equipment for measurement, control and laboratory use'*	
Compiled by (+ signature): Checked by (+ signature): Date of issue:	
Evaluated by:	*If different to 'Compiled by'*
Address:	*Give here the address of the company or individual who carried out the evaluation*
Type of equipment:	*Indicate the product category, e.g. Portable Hi-Fi, Radio/Cassette*
Model/type reference:	*Indicate the product's exact model name – this should be the exact type number as indicated on the model name (rating) label*
Manufacturer's name and address:	*Indicate here the manufacturer's name and address*
Operating condition:	*Indicate whether the equipment operates in continuous, short-term, intermittent mode, etc.*
Equipment mobility:	*Indicate if the equipment is movable, handheld, stationary, fixed, permanently connected, etc.*
Size (H x W x D):	*Indicate the size of the equipment*
Mass (kg):	*Indicate the equipment's weight (kg or g)*
Phase (single/three):	*Indicate if the equipment is operated from a single or a three-phase supply*
Protection against ingress of water:	*Indicate if precautions have been taken to protect the equipment from ingress of water – and, if so, how*
Classification:	*Indicate whether the equipment is of Class I, II or III construction*
Current rating:	*Indicate the current rating (in A, mA, etc.)*

Report No: 001   Page 2 of 21

Voltage rating (range):	*Indicate the operating voltage or voltage range*
Power rating:	*Indicate the power consumption (in W, W/h, kW, etc.)*
Environment of use:	*Indicate the environment of use, e.g. light industrial*
Connection to the supply mains:	*Indicate the method of connection to the mains supply*
Copy of marking plate and other labels:	

CONCLUSIONS:

Conclude in this section whether you believe the product has been adequately evaluated and that the essential safety requirements have been met

Report No: 001 **Page 3 of 21**

Section 1

Section	Requirement – Test	Remarks	Results
1.1.0	MARKING and INSTRUCTIONS	Sections 1.1.0 to 1.1.11 refer to the rating label	
1.1.1	Legibility	Confirmation that the label provides legible and clearly discernible information	
1.1.2	Durability	Explanation of how the label was tested to confirm durability.	
1.1.3	Rated voltage (V)	e.g. 110–240 V	
1.1.4	Rated power (W or VA)	e.g. 88 W	
1.1.5	Rated current (A)	e.g. 1 A	
1.1.6	Symbol for nature of supply	e.g. ~	
1.1.7	Rated frequency (Hz)	e.g. 50 Hz	
1.1.8	Symbol for Class II (if applicable)	e.g. The double square mark.	
1.1.9	Equipment manufacturer	e.g. Ascot Manufacturing Company.	
1.1.10	Type/model	e.g. AM-601L	
1.1.11	Safety information	Evaluation of warnings and labels, e.g. warning of the presence of high voltage, the need for earthing, etc.	
1.1.12	Fuse holder	Confirmation that the fuse type is indicated on the product, e.g. a label is placed adjacent to the fuse holder or marking on the PWB indicating type used, rating, voltage, etc.	
1.1.13	Mains switch	Confirmation of how the mains switch is marked, e.g. 'on'/'off', 'O'/'1', etc.	

Report No: 001   Page 4 of 21

1.1.14	Earth symbol	Confirmation that a label with a symbol in accordance with IEC 417 is placed adjacent to the protective earth terminal.	
1.1.15	Marking for voltage/frequency settings	If a selector switch is used, confirmation that the voltage and frequency settings are clearly marked adjacent to it.	
1.1.17	Placing of markings	Indication of the position of markings and labels, e.g. should be placed adjacent to controls, terminals, etc.	
1.1.18	Instructions when protection relies on building installation	Confirmation that warnings are used, e.g. the need to connect through a protective device such as an isolator, RCD, etc.	
1.1.19	Letter or symbols used (according to IEC 417)	Confirmation that symbols used comply with IEC 417.	
1.1.20	Instructions for use – language	Evaluation of the user instructions, e.g. are they given in the language of the user?	
1.1.21	Exclamation mark in triangle ⚠	Confirmation that the exclamation mark is indicated in the service manuals (this mark indicates that Safety Critical parts need be replaced by identical components).	

Note: *The markings and labels fitted on the equipment under evaluation must be assessed for contents, durability, positioning, etc., according to the individual standard's requirements. It is also advisable to use as many warning labels as necessary, and make clear, accurate and unambiguous cautions and instructions in the user manual. Also refer to: 10.16 Labels, 10.17 Markings, 10.18 User instructions.*

Report No: 001　　　　　　　　　　　　　　　　　　　Page 5 of 21

Section 2

Section	Requirement – Test	Remarks	Results
2.1.0	PROTECTION AGAINST HEATING HAZARDS – under NORMAL operating conditions	An explanation could be given here of how temperature measurements were taken, e.g. after x hrs' continuous operation at x V input voltage, under certain loading, etc. See Table 1	
2.1.2	Temperature rise on accessible parts	Explanation of where measurements were taken and confirmation that temperature rises met the requirements of the applicable standards. Refer to Table 1.	
2.1.3	Temperature rise on parts providing electrical isolation	As above	
2.1.4	Temperature rise on parts acting as mechanical barriers	As above	
2.1.5	Other	As above	

Section 3

Section	Requirement – Test	Remarks	Results
3.1.0	PROTECTION AGAINST HEATING/OTHER HAZARDS – under FAULT operating conditions	An explanation could be given here of how and where the temperature measurements were taken (while the fault was applied). See Table 2	
3.1.1	Shock hazard	Explanation of what accessible conductive parts were checked and whether they were found to be safe during the application of the fault.	

Report No: 001

3.1.2	Hazard from softening solder	*Confirmation that softening of solder was not evident.*	
3.1.3	Measurement of temperature rise	*Confirmation and listing of temperature measurements taken in areas and components of the circuitry. Refer to Table 2. Confirm that temperature rises measured were in accordance with the maximum permissible limits allowed by the product standard.*	
3.1.4	Temperature rise on accessible parts	*Confirmation that temperature rises on accessible parts were safe and in accordance with the applicable standard.*	
3.1.5	Temperature rise on PWBs	*Confirmation that temperature rises on PWBs were found to be in accordance with the applicable standard.*	
3.1.6	Blocked fan rotors, vents, paper jam, etc.	*Mention any other fault conditions introduced and results obtained.*	

Report No: 001

Section 4.1

Section	Requirement – Test	Remarks	Results
4.1.0	PROTECTION AGAINST ELECTRIC SHOCK	See Tables 3, 4 and 5	
4.1.1	Protection against operator contact	Confirmation that hazardous voltage wires are insulated and that no bare wires are exposed to the operator.	
4.1.2	Internal wires	Confirmation that no internal wires are accessible to the user.	
4.1.3	Service area	Confirmation that unintentional contact is unlikely during service operations, and sufficient warning is placed in critical areas.	
4.1.4	Shafts, knobs, handles, etc.	Confirmation that metallic handles, knobs, etc. are not likely to become live.	
4.1.5	Pins and plugs	Confirm that tests have been performed to ensure that the mains plug is safe after its withdrawal from the socket.	
4.1.6	Access via ventilation holes	Confirm that tests were performed to assess user accessibility to live parts via the ventilation holes.	
4.1.7	Terminal devices	Confirm that tests were performed to assess user accessibility to live parts via the terminal devices.	
4.1.8	Pre-set controls	Confirm that tests were performed to assess user accessibility to live parts via the pre-set controls.	
4.1.9	Removal of protective covers by hand	Confirmation that the removal of protective parts is only possible by the use of a tool.	

Report No: 001					Page 8 of 21

4.1.10	Measurement of peak current	If applicable, explanation of how the measurement is performed and results obtained.	
4.1.11	Method of insulation	Explanation of how dangerous voltage insulation within the product is achieved, e.g. by the use of solid or laminated materials of adequate thickness, adequate creepage distances are maintained, etc.	
4.1.12	Insulating materials	Confirmation that materials such as natural rubber, materials containing asbestos or hygroscopic materials are not used as insulators.	
4.1.13	Dielectric strength test	Reference to Table 3 should be made here.	
4.1.14	Insulation resistance test	Reference to Table 4 should be made here.	
4.1.15	Bridging of double or reinforced insulation	Confirmation that only approved components were used to bridge the primary to secondary circuit insulation.	
4.1.16	Detaching of wires	Confirmation that wires likely to become loose and touch primary live components are secured and that safe creepage and clearance distances are maintained.	

Report No: 001   **Page 9 of 21**

Section 4.2

Section	Requirement – Test	Remarks	Results
4.2.0	PRIMARY POWER ISOLATION (requirement by some standards)		
4.2.1	Disconnection device	Explanation of how the equipment is disconnected from the mains supply, what device is used and whether it is of an approved type.	
4.2.2	Isolator used	If an isolator fixed to the building is used to isolate the mains input power – confirmation of type and approval status.	
4.2.3	Disconnection of both poles simultaneously/all poles, etc.	Confirmation that both poles of the supply or all phases are disconnected when the equipment is switched in the 'off' position.	
4.2.4	Marking of switch/isolator	Confirmation of such markings and their location.	

Section 4.3

Section	Requirement – Test	Remarks	Results
4.3.0	OVERCURRENT AND EARTH FAULT PROTECTION IN PRIMARY CIRCUITS		
4.3.1	Number and location of protective devices	*If more than one protective device is used, indicate the number, type and location.*	
4.3.2	Type of protective device and breaking capacity	*If a protection device is used, confirmation of the type, its approval status and whether it has an adequate breaking capacity.*	

Section 4.4

Section	Requirement – Test	Remarks	Results
4.4.0	PROTECTIVE EARTH		
4.4.1	Reliable connection	*Confirmation that all accessible metal parts are earthed, looped and connected to the mains earth, etc.*	
4.4.2	Green/Yellow insulation	*Confirmation that all insulated earth conductors are green/yellow.*	
4.4.3	Risk of corrosion	*Confirmation that there is no risk of corrosion, i.e. by using the same metals in the earth connection system.*	
4.4.4	Earth connector resistance	*Confirmation of tests carried out between the appliance earthing pin in the AC inlet and accessible metal parts – indicate the applied current, the measured resistance and whether it meets the applicable standards criteria.*	

Report No: 001　　　　　　　　　　　　　　　　　　Page 11 of 21

Section 4.5

Section	Requirement – Test	Remarks	Results
4.5.0	EARTH LEAKAGE CURRENT (required by some standards)	See Table 6	
4.5.1	Test voltage applied	Indication of applied voltage, e.g. test carried out at mains voltage $240V + 10\%$.	
4.5.2	Measured current (mA)	Indication of measured current and confirmation whether within the limits of the applicable standard. Make reference to Table 6.	

Section 4.6

Section	Requirement – Test	Remarks	Results
4.6.0	WIRING, CONNECTION TO SUPPLY		
4.6.1	Type of connection	Provide indication of how the equipment is connected to supply mains, e.g. supply is provided by means of a flexible (2.5 m) power cord – fitted with approved connectors type ... etc.	
4.6.2	AC inlet	If an AC inlet is used, confirmation of the type used, its approval status, etc.	
4.6.3	Construction and fixing of mains terminal	Detail the construction and fixing of the mains terminal, e.g. is it connected directly from the power cord or by using an AC inlet?	

Report No: 001   **Page 12 of 21**

4.6.4	Power cord and cross-sectional area	*Confirmation of the power cord's cross-sectional area and whether it is adequate for the power requirements of the product.*	
4.6.5	Cord securing	*Confirmation of how the power cord is secured to the product and indication of tests performed to confirm that no displacement will occur when harshly pulled.*	
4.6.6	Risk of cord damage by bushing	*If a cord bushing is used, confirmation that there is no risk of cord damage.*	

Note: *In Chapter 10 the reader can find details of temperature measurements and methods of confirming rises under normal operation and under fault conditions. Maximum permissible temperature rises for the materials used in the construction of the product (e.g. for the cabinet, PWBs, etc.) as required by the applicable standard or regulation, must be known by the designer or the manufacturer before testing begins. The measured value must be below the specified limit in the standard before it can be assumed that the product is safe. One of the most important safety requirements is that both the operator* and *service personnel are protected from electric shock. The isolation of the primary input power, the integrity of the safety earth and the disconnection of the power supply during a fault condition all need to be assessed and their adequacy confirmed.*

Report No: 001   Page 13 of 21

Section 5.1

Section	Requirement – Test	Remarks	Results
5.1.0	**PROTECTION AGAINST MECHANICAL HAZARDS**		
5.1.1	Vibration/Bump/ Drop test	*If applicable, decribe what tests have been performed, the parameters used and how compliance has been confirmed.*	
5.1.2	Impact test	*If applicable, confirmation should be given that external areas have been assessed and that live parts are protected and did not become accessible during or after the test.*	
5.1.3	Fixing of knobs, handles, etc.	*Confirmation that knobs, handles, etc. are secured and that if removed live parts will not become accessible.*	
5.1.4	Stability test (required by some standards only)	*Confirmation that free-standing equipment is stable and will not overbalance and cause injury to the user, e.g. a TV falling on a child.*	
5.1.5	Edges and corners	*Confirmation that there are no edges or corners on the equipment which represent a hazard to the operator.*	

Section 5.2

Section	Requirement – Test	Remarks	Results
5.2.0	**ELECTRICAL CONNECTIONS AND MECHANICAL FIXINGS**		

Report No: 001

5.2.1	Cover fixings	Description of how the cover is fixed, e.g. all covers were fixed by screws (10 mm × 2 mm) – ensure that creepage distance and clearances are maintained.	
5.2.2	Fixing of the enclosure	Description of how the enclosure is fixed – e.g. the enclosure is fixed by screws (10 mm × 2 mm) – ensure that creepage distance and clearances are maintained.	
5.2.3	Fixing of internal wiring	Confirmation that all internal wiring is secured using cable ties, holders, etc., maintaining safe creepage and clearance distances.	
5.2.4	Fixing of uninsulated conductors	Confirmation that uninsulated wires are secured, e.g. by the use of cable ties, etc. – ensuring safe creepage and clearance distances are maintained.	

Section 6

Section	Requirement – Test	Remarks	Results
6.1.0	COMPONENTS	See Table 7	
6.1.1	Use of approved safety critical components	Confirmation that all safety critical components used in the product are approved. Make reference to Table 7 for details.	
6.1.2	Tests on non-approved safety critical components	Identify any non-approved Safety Critical parts used and describe what additional tests were carried out to ensure their safe operation.	

Report No: 001   Page 15 of 21

Section 7

Section	Requirement – Test	Remarks	Results
7.1.0	PROTECTION AGAINST FIRE HAZARDS	See Table 8	
7.1.1	Flammability of: • Materials • Components • Wiring harness • Enclosure • Decorative parts • PWB • etc.	Confirmation that relevant parts and materials of suitable flammability rating have been employed – ratings should be as specified by the appropriate standard. Make reference to Table 8.	
7.1.2	Protection against spread of flammable liquids	Confirmation that flammable liquids present within the equipment are suitably contained.	
7.1.3	Flammability of high voltage components	Confirmation that high voltage components used are supplied with the manufacturer's certification for flammability, test data or have been internally tested.	

Section 8

Section	Requirement – Test	Remarks	Results
8.1.0	PROTECTION AGAINST SPLASHING WATER		
8.1.1	Use of appropriate enclosure	Confirmation of the use of an appropriate enclosure if the equipment is susceptible to splashing water.	
8.1.2	Splash treatment	Confirmation of tests performed to ensure safe operation when water is splashed on the product.	

8.1.3	Humidity treatment	Confirmation of the evaluation for safe operation in a high humidity environment.	

Section 9

Section	Requirement – Test	Remarks	Results
9.1.0	PROTECTION FROM OTHER HAZARDS		

9.1.1	• Implosion
	• Chemical
	• Energy
	• Radiation:
	  – X-ray
	  – laser
	  – ultraviolet
	  – microwave
	• Acoustic
	• Overflow
	• Spillage
	• Liberation of poisonous gases
	• Explosion
	• etc.
		Confirmation of evaluations and actions taken to eliminate operator risk from any other hazards.	

Note: *All sources of hazards that could result in injury to the operator or service personnel must be considered and dealt with. These are specific to the product, its relevant standard and the environment of use. The manufacturer needs to explain in his report how he has addressed these hazards and what actions have been taken to minimize the risks associated with their presence. Detailed explanation must therefore be presented in the report together with calculations, lists of appropriate components, etc.*

Report No: 001

Section 10 – Tables

Table 1 Temperature rise (dT) measurement (normal condition)

Operating conditions: e.g. Normal operation – downloading data to a printer, etc.
Ambient temperature: e.g. 21°C

Part/location of temperature rise:	Test voltage	dT measured (°C)	Test voltage	dT measured (°C)	dT limit (C)
e.g. D 601 – 5V regulator	198 V	50	254 V	54	85
e.g. Enclosure – rear	198 V	15	254 V	15	60

Table 2 Temperature rise (dT) measurement (fault condition)

Ambient temperature: e.g. 21°C

Component	Fault	Test voltage	Monitor point	Results (dT)
e.g. Capacitor 101	Short circuit	198 V (i.e. −10%)	PWB coil 601	*Temperature rise of 55°C in less than 10 seconds before power supply shutdown.*
e.g. Transistor M621 base/emitter	Open circuit	254.4 V (i.e. +6%)	PWB capacitor 500	*Immediate shutdown of power supply. Fuse blown.*

Report No: 21   Page 18 of 21

Table 3 Dielectric strength test

Test voltage applied between:	Test voltage and test duration	Result (i.e. breakdown of insulation)
e.g. Mains poles (live–neutral) – primary fuse disconnected	x V pk y sec	
e.g. Mains poles connected together and in/out terminal	x V pk y sec	
e.g. Mains poles connected together and operator metal accessible parts (metal enclosure)	x V pk y sec	

Table 4 Insulation resistance

Resistance measured between:	Applied voltage and test duration	Result (Ω)	Limit (Ω)
e.g. Mains poles (live–neutral) – primary fuse disconnected	x V DC y sec		
e.g. Mains poles connected together and in/out terminal	x V DC y sec		
e.g. Mains poles connected together and operator metal accessible parts (metal enclosure)	x V DC y sec		

Report No: 001

Table 5 Clearances (cl) and creepage (cr) distances

Pollution degree: e.g. *III*

Distance measured between:	cl measured	cl required	cr measured	cr required	Result (pass/fail)
e.g. Live parts and safety earthed parts – Class I	x mm	y mm	x mm	y mm	
e.g. Mains transformer primary and secondary	x mm	y mm	x mm	y mm	
e.g. Mains transformer primary and core	x mm	y mm	x mm	y mm	
e.g. Mains transformer secondary and core earth	x mm	y mm	x mm	y mm	
etc.					

Report No: 001 **Page 20 of 21**

Table 6 Earth leakage current

Operating condition	Test voltage	Measured current	Limit and result (pass/fail)
e.g. Normal–full load connected	254.4 V (i.e. +6%)	Phase 1 (live): mA Phase 2 (neutral): mA	mA
e.g. Stand-by – partial load, etc.	254.4 V (i.e. +6%)	Phase 1 (live): mA Phase 2 (neutral): mA	mA
etc.			

Table 7 List of Safety Critical components

Component	Reference number	Type or model	Manufacturer	Rating	Standard	Approval
e.g. Mains switch	SW1	KSD-3FF	SENKO	4A/120A	BS ...	BSI/S/D/VDE
e.g. Fuse	F100	215	Ampex	T4AH 250V	EN 60127	BSI KM 7670

Table 8 List of flame retardant materials

Material or component, etc.	Flammability rating	Acceptable
e.g. Enclosure	94-V5	Yes
e.g. Wire between RCD1 and Transformer 2	VW-1	Yes
e.g. PWB, etc.		

Report No: 001   Page 21 of 21

Section 11 – Photographs

It is advisable to photograph the equipment under evaluation and attach prints to the evaluation report. Ideally, photographs should be taken of:

- the front and rear of the equipment
- internal construction
- wire routing close-ups
- PWBs (showing Safety Critical parts)
- warning, cautionary and rating labels
- etc.

11.4 Summary

Although external evaluation by accredited laboratories is clearly the best option to ensure that legal obligations are satisfied, it is recognized that this is not always possible for many small manufacturers.

As an alternative therefore internal self-evaluation is a very sensible and responsible approach which a manufacturer can take in order to confidently apply the CE Mark to new products.

The example report given in this chapter can be of use to most such organizations, but it must be appreciated that it is just an example. The test and evaluation parameters suggested offer general guidelines only, and the report will need to be customized in order to ensure the satisfactory and comprehensive evaluation of any one particular product or product category – in all cases the product specific standard should be consulted to determine the appropriate and relevant limits.

Appendix 20 explains how a blank copy of the suggested report format may be obtained and modified to enable its use for particular product evaluation.

Part 3

Preparing the documentation

Chapter 12

Technical File

The technical documentation (Technical File) provides a means of assessing whether the electrical equipment conforms with the requirements of the Low Voltage Directive. It must include details of the design, manufacture and operation of the electrical equipment in so far as these details are needed to assess the conformity of the electrical equipment with the requirements of the Directive.

The Technical File should be a comprehensive document and must include:

- general description of the electrical equipment
  - a copy of the user instruction manual containing model name, etc. should be sufficient to satisfy this requirement
- conceptual designs and manufacturing drawings, etc.
  - a copy of mechanical and electrical drawings, block and interconnecting diagrams
  - lists of parts and Safety Critical components (including approvals status/marks, etc.)
  - a list of Safety Critical assemblies (assemblies that have Safety Critical significance)
  - copies of Safety Critical operation standards
  - description of design concepts which have been followed, e.g. power supply operation, etc.
  - explanation necessary for understanding the circuit diagrams (a copy of the service manual might be sufficient)
- results of design calculations made, examinations carried out, etc.
  - copies of evaluations and 'in-house' test reports
  - open- and short-circuit test results
  - copies of internal and external investigations
- test reports
  - including copies of internal or external test reports prepared to demonstrate compliance with a Harmonized, International or National Standard or Regulation.

This report could be prepared by the manufacturer, a Notified Body or any other person the manufacturer considers to be competent.

- declaration of conformity
  - a copy of the Declaration of Conformity as given in Chapter 13
- other
  - a list of alternative Safety Critical parts
  - a copy of the procedure employed to control manufacturing changes to Safety Critical components and assemblies
  - copies of procedures followed to:
    (a) handle end-of-line failures (i.e. marking, investigation, etc.)
    (b) perform internal and external calibrations of Safety Critical equipment
    (c) perform routine safety testing (on or off the assembly lines)
    (d) inspect incoming components
    (e) control Certificates of Conformance or Declarations of Conformity
    (f) perform quality checks (a copy of BS 5750/ISO 9000 procedures)
  - risk assessment calculations
  - copies of delivery and commissioning instructions
  - lists of limitations of use.

For easy understanding, flowcharts illustrating the above procedures should be provided wherever possible (rather than long documents).

In large organizations where design, parts control and production are situated in different parts of the factory (or even in entirely different plants), the Technical File may be located in any one of these areas or possibly certain parts of it in each one. A control procedure will therefore need to be set up to ensure that regular auditing of these locations is carried out by responsible personnel. All such audits must of course be documented.

A summarized list of the Technical File contents is given below.

The Technical File and other safety related documents (for details refer to Section 14.9) should be kept for a minimum period of 10 years after the last production. It is recommended that two copies are kept in separate locations. Documents stored only on disk or by other electronic means should be kept in a fireproof cabinet.

Technical File contents

Section 1: Company CE Policy
Section 2: Declaration(s) of Conformity
Section 3: Descriptions of products tested
Section 4: Description of chassis design
Section 5: Comparison tables between products covered
Section 6: Mechanical drawings
Section 7: Block and schematic diagrams
Section 8: Calculations/tests results
Section 9: External test/evaluation reports
Section 10: Safety Critical components and other parts lists
Section 11: Instructions for operations and limitations
Section 12: Instructions for delivery/installation/maintenance
Section 13: Copies of SC parts list/D of C/C of C
Section 14: End-of-line test procedures
Section 15: Quality Assurance test procedures
Section 16: Production control procedures
Section 17: Change control procedures
Section 18: Risk assessment
Section 19: Other

Chapter 13

Declaration of Conformity

The Declaration of Conformity is a written declaration by the manufacturer or its authorized representative, that the equipment to which the CE Marking has been affixed complies with the requirements of the LVD. It is a statement to confirm that the product meets all the requirements of the LVD and indicates how they were satisfied.

13.1 Declaration of Conformity – contents

The EC Declaration of Conformity must contain the following information:

- The name and address of the manufacturer or his authorized representative established in the Community.
- A description of the electrical equipment.
  A clear identification of the equipment must be made on the Declaration. It is recommended that a full description of the equipment be given. This should include the model type, the model number, etc.
- References to the Harmonized standards which have been applied, citing relevant clauses or exceptions (if applicable).
  If a Harmonized standard has not been applied, references to the International or National standards which have been applied must be listed.
  If no standards have been applied a clear indication of the method employed to meet the safety objectives listed in the LVD must be included here.
  It is also useful to refer any other regulations which are applicable to the product. If a regulation is in transition, it is advisable to outline whether or not the requirements of that regulation have been applied.
- Identification of the signatory who has been empowered to enter into commitments on behalf of the manufacturer or his authorized representative established within the Community.

This should identify an individual within the manufacturer's (or his authorized representative's) organization.
- The Declaration of Conformity must be drawn up in one of the official languages of the Community.

A copy of the EC Declaration of Conformity is not required to accompany each individual product sold, but a copy must be kept within the EU territory by the manufacturer, the authorized representative, or the person who first placed the equipment on the EU market. A copy of the Declaration of Conformity must also be kept with the Technical File.

If there are reasonable grounds for suspecting that the product is unsafe, the enforcement authorities may request a copy of the Declaration of Conformity and other associated documentation. Failure to make the documentation available within a reasonable time constitutes an offence.

In making a Declaration of Conformity, not only is the company accepting responsibility, but also the signatory becomes *personally* liable, and runs the risk of a prison sentence in the most extreme cases. The declaration applies to all production items and a good quality monitoring process is essential.

Examples of a Declaration of Conformity are shown in Appendix 13.

Part 4

Setting up production control

Part II

Sampling and production control

Chapter 14

Production control

To ensure continued conformity, the manufacturing quality system will need to include control of operations including process inspection and testing at certain stages of the manufacturing process. The level of inspection and testing will differ between manufacturers and will depend on the complexity of the product, the technology employed, the range of variants, the volume of production and other factors.

Areas of the manufacturing process that are critical in ensuring that the product is manufactured in a controlled manner must be identified and targeted for action, otherwise *unsafe* products will be produced every time.

A production control system should be easy to set up, it should also be well documented, and based on the three key areas of:

- control of components and documentation (see Section 14.3);
- inspections at key manufacturing stages;
- testing before the despatch of the product.

$$Control + Inspection + Testing = Production\ control$$

14.1 Essential 'tools'

The procedure of setting up a production control system should begin by identifying key areas of the process and providing them with the necessary 'tools' to fulfil their function. As a first step there will be a need to:

- identify and list all Safety Critical components
- identify and list all Safety Critical operations
- prepare and issue operation standards (individual work instructions)
- prepare test procedures and manuals.

Following this, an operator training and safety awareness programme will need to be implemented.

14.1.1 Safety Critical components lists

Some of the components used in the assembly of a product are critical for its safe operation. Parts in the primary circuit (the 240 V AC area), parts crossing the 'barrier' between primary (high voltage) and secondary (low voltage) areas, and parts between the primary circuit and the user are usually classed as *Safety Critical*. Externally prepared test reports of the product will include a list of such components.

Compiling a list of Safety Critical components is an essential task in any manufacturing plant as it will be used in all areas where component identification or inspection is necessary, i.e. incoming goods, material stores, production lines and quality control.

The list can be as simple or as comprehensive as the manufacturer wishes. An example of a typical Safety Critical component list is given in Table 14.1.

Table 14.1 Example of a Safety Critical components list

Safety Critical Components List

Part description	Value or rating	Identification	Model/Location	Part number	Supplier	Approval mark
Capacitor	2200 pF	NS220	SV-335/C233	451-231	HSR	UL, CSA S, D, N
Isolator	16A	DS3-FF	etc.			
Relay	...					
etc.						
Issued by:		Issue date:		Issue no:		

14.1.2 Safety Critical assemblies

Production line operators are responsible for carrying out tasks such as wire routing, component placement and soldering, and many other assembly operations that could have significant safety implications. Wire routing is often critical, especially in cases where a safe distance has to be maintained, e.g. in cases where single insulation has been used on high voltage wiring.

A list identifying such *Safety Critical operations* should be compiled and made available to the engineers who write the operation standards.

The same list will be used by quality control personnel in their detailed inspection of finished product samples. An example of how such a list may be laid out is given in Table 14.2.

Table 14.2 Example of a Safety Critical assembly list

Safety Critical Assembly List

Task	Model	Comment
Securing of the earth conductor	KD3-N	Ensure the star washer is fitted over the earth stud first, then place the ring connector on top followed by a lock washer and tighten.
Affixing the safety earth label	KD3-N	Ensure it is placed adjacent to the earth stud.
Wire routing of N3–D3 lead		
etc.		
Issued by:	Issue no:	Issue date:

Table 14.3 Example of an Operation Standard

OPERATION STANDARD		Issue no:		
TITLE: e.g. Fitting of the earth label	Issued by:	Model:	Date:
'Explanation' (explain the task in detail)	Caution: This is a Safety Critical Operation
	'Attach a photograph or drawing'

14.1.3 Preparing operation standards

Industrial engineers or other personnel creating operation standards would find the Safety Critical components and assembly lists invaluable. Clear identification on the operation standards of critical components and critical operations through the use of drawings or photographs combined with focused operator training would ensure that all products are made identically and in accordance with the initially tested sample. Operation standards should be authorized by a responsible engineer, given an issue number and dated – an example is given in Table 14.3.

14.1.4 Preparing a test manual

Manufacturers producing goods in large volume may need to consider some additional testing where every day at least one unit from each assembly line is tested for confirmation of its ongoing compliance with the applicable standard. A manual would need to be prepared describing the detailed inspection procedures and test methods.

It should be prepared either by the designer, the quality control section or the engineer responsible for approvals. The tests prescribed should be based on the safety testing requirements specified in the relevant standard, but may necessarily be simplified or curtailed as probably not all specialized equipment or tools would be available.

14.2 Production control procedures

Production control =

Control	Of assembly operations (e.g. by following operation standards)
	Of purchasing of new/alternative parts (e.g. by using the Safety Critical parts list)
	Of engineering change (e.g. by using the Safety Critical parts list)
+	
Inspection	At QC area or process control (e.g. by using the Safety Critical assembly list)
	At assembly areas (e.g. by auditing adherence to operation standards)
	At incoming goods area (e.g. by using the Safety Critical parts list)
+	
Testing	Periodically to a sample plan at QC area (e.g. by using the test manual)
	Every product before despatch (e.g. by following operation standards)

14.3 Engineering change control

Engineering changes are inevitable during a product's lifetime for a number of reasons, e.g. alternative parts supplier, change of performance criteria, market requests, etc. The majority of such changes will have no impact on safety. Nevertheless, a clear procedure need to be established which differentiates between the 'harmless' changes and those which require evaluation.

In order to make correct decisions, a good understanding of the safety principles and the standards applicable to the product is necessary. Organizations which produce large ranges of products in large quantities may have to submit proposals for technical (engineering) changes to their respective external accreditation bodies (if initial certification was so gained) for approval. This can be a tedious and time consuming task.

A scheme offered by most certification agencies and followed by many large manufacturers is to have an in-house specialist engineer, authorized by the particular certification body, to control and approve (by signature) safety related changes. A similar system could be set up by any manufacturer, where only one or a few designated persons are trained and authorized to approve safety related changes. Ideally these would be engineers in the quality control or design sections.

After the product has been tested and released to production, subsequent change requests are usually originated by Design or the Production Engineering teams. Such change requests must be correctly controlled and documented; they should also be presented in a manner which should not pose a risk of ambiguity.

Change request forms should be designed carefully so as to give every detail necessary to guarantee accurate information transfer and correct introduction of the change.

Engineering changes can be of two alternative types:

- Engineering Change Note (ECN)
- Production Change Request (PCR).

ECNs are used when the originator wishes to implement a component change to the product in manufacturing. A form as per the example shown in Appendix 15 could be used. The change note should give details of:

- the change
- the part or component's reference number
- the part or component's type/part number, etc.
- the starting and finishing accumulation of the change
- the models or products affected
- etc.

ECNs must be signed and dated by the originator, authorized by his superior and the Purchasing Department, and approved by the person responsible for Safety & EMC product approval (or his deputy).

A PCR is used in cases where the designer wishes the production section to apply an operational or assembly change, e.g.

- to the wiring dressing routing
- to the torque level of certain screws
- assembly sequence change
- etc.

Again, a form should be raised instructing the relevant sections to apply the change. As with the ECN, it must be authorized and approved (especially by the person responsible for Safety & EMC product approval) before implementation. Details of the change should be supplied to the relevant sections/persons in detail – ideally supported by photographs or diagrams (see Appendices 15 and 16).

14.4 Control of purchasing

One of the most important product safety considerations is control over the purchase of parts. Purchasing officers need to be fully aware of the parts listed as *Safety Critical*, and those that are acceptable alternatives.

In cases where the purchase of alternative components is unavoidable, then consultation with the designers and the quality control department will be necessary – a procedure should be available for dealing with such a situation.

14.5 Confirmation at incoming goods area

Using the Safety Critical components list, a confirmation check must take place at the incoming goods area and only approved components should be accepted (where specified). In cases where the consignment contains non-approved parts, these must be segregated immediately (in a dedicated quarantine area) until an investigation has been completed.

14.6 Adherence to operation standards

Clear operation standards containing detailed instructions are necessary if the assembly operators are to produce a product of repeatable safety and quality performance. All operations, whether permanent or temporary, must be covered by an appropriate operation standard.

When preparing operation standards, the following must be considered:

- assembly instructions must be clear and concise
- photographs or drawings must be consistent with the 'prototype' which was originally evaluated for safety approval
- Safety Critical components and operations must be clearly identified

- tools to be used are fit for purpose (specify critical factors such as torque settings).

14.7 Routine testing

In addition to the control of parts and the confirmation of assemblies in the incoming goods and production areas respectively, a manufacturer must perform some testing to ensure that every product made is in conformity with the applicable standard or regulation.

Routine tests are usually performed away from the production area, probably in the quality control department. The operator should follow the procedures detailed in a test manual and record the test results. Routine tests and inspections should include:

- a check of the correct AC input wire polarity and connection method of sub-assemblies
- the correct value of Safety Critical components
- the protective earth connection of screen and metal barriers (applicable to Class I products only)
- the use of the correct power cord, its assembly and its anchorage
- the correct wire routing within the equipment
- the correct fit of internal plug connections
- the correct safety relevant marking and labelling inside and outside the equipment
- the correct mounting of mechanical parts and guards
- access to hazardous voltage areas via openings
- the performance of some special tests, such as X-ray emission, etc.

14.8 Testing at the end of line (final inspection)

Safety testing at the end of the production line, should be carried out on a 100% basis – every product despatched must be guaranteed as being a safe product. This test is essential to identify possible errors caused by manual or even automated operations on the production line. It should be performed just prior to packing, with all covers and guards in place and the enclosure fitted.

End-of-line safety testing on a sample basis is not advisable – just one failure in the market could be too costly! Even if the manufacturing process is perfect, there might be occasions where evidence is necessary to demonstrate that a particular equipment was indeed tested and found to comply.

The test must be carried out according to an appropriate procedure stating the relevant test parameters such as voltage, current, duration, etc. Test equipment must be regularly confirmed – daily for flow-line production, or just before use in the case of small-scale, infrequent production runs. The test equipment must be fully calibrated on at least a 6-monthly or annual basis. This calibration could be performed internally or externally – depending on the availability of a suitable standard for comparison.

All operation standards, test procedures, calibration certificates and other records pertaining to relevant test equipment, together with records of serial numbers of products tested, should be kept for as long as practicable after production of a particular model has ceased.

Four types of final safety tests are quoted in most standards:

- Earth continuity
- Dielectric strength
- Insulation resistance
- Earth leakage current.

14.8.1 Earth continuity

For Class I equipment, i.e. those supplied with a protective earth, the earth continuity should be checked between the protective earth contact of the mains plug, or appliance inlet, or the protective earth terminal (in case of a permanently connected equipment) and:

- the accessible conductive parts, including terminals regarded as accessible which should be connected to the protective earth terminal *and*
- the protective earth contact of socket-outlets (if provided to deliver power to other equipment).

The test current for each product category may vary from 1.5 times the fuse rating to 25 A. It should be derived from a source having a no-load voltage of about 12 V and should be applied for 1 to 4 seconds – usually specified to be less than 0.1 Ω although some standards specify $0.5\,\Omega + R$ (R: resistance of power cord); 0.5 Ω is also acceptable for the lighting industry.

Equipment covered by the Machinery Directive has a special test defined in EN 60204. The protective wire is first checked visually and manually by manipulation. Then by applying 10 A (AC) for at least 10 seconds to several points of the protective earth, the voltage drop is measured over the full length of the protective wire. The voltage drop could vary from 3.3 V to 1 V for 1.0 mm to >6.0 mm (at the smallest effective cross-section of the protective wire for the branch tested).

14.8.2 Dielectric strength (flash) test

To prevent an insulation fault as a result of incorrect assembly or component breakdown, the insulation of the equipment should be checked by the following tests.

Table 14.4 Dielectric strength test – values based on consumer product standards

Application of test voltage to	Test voltage
Basic insulation	2120 V peak (1500 V rms) No flash-over should occur during the test
Reinforced insulation	4240 V peak (3000 V rms) Trip current 6 mA (8 mA peak)
Test duration	6 sec for manual or 2 sec for automated test equipment

An AC test voltage of a substantially (mains frequency) sine-wave form, or a DC test voltage, or a combination of both with peak value as specified in Table 14.4 should be applied between the mains supply connection and:

- terminals regarded as accessible and
- accessible conductive parts.

During the test, mains and functional switches should be placed in the 'on' position and the voltage should be applied for a minimum of 1 second, although some standards specify 10 seconds (allowing for ramp times). No flash-over or breakdown should occur during the test.

The test source is usually provided with a current sensing (overcurrent) device which, when activated gives a 'pass' or 'fail' indication.

For some standards the trip current limit is 100 mA – tripping at this current should be regarded as flash-over, i.e. 'fail'.

14.8.3 Insulation resistance

The insulation resistance is to be measured between both poles of the mains supply cord or terminals (connected together) and all accessible metal parts. The resistance is measured at voltages of between 500 V and 2500 V (DC). The minimum measured value for some products should be 2 MΩ, while for others the standards specify 7 MΩ.

14.8.4 Earth leakage current

For some products, such as information technology (EN 60950), medical (EN 60601) and laboratory equipment (EN 61010), an earth leakage current test is specified (note: for EN 60950 it is required in place of the insulation resistance test). This test must be carried out at the rated mains input voltage (although some standards ask for this to be increased by between 6% and 10%), and the leakage current is measured via the protective earth. The maximum permitted current will vary according to the relevant standard but (e.g.) for ITE is:

- Class I products: 3.5 mA.
- Class II products: 0.25 mA.

Note: *All values for the end-of-line test quoted above are for reference only. Each product specific standard quotes particular requirements, often differing between compliance and end-of-line testing. The relevant standard should always be consulted.*

14.9 Records

All procedures and operations related to any safety testing should be documented. A simple document identifying the systems in place and their purpose, would be very useful in helping an enforcement inspector to understand and evaluate a manufacturer's operation.

Records of all tests and confirmations should be kept for a minimum period of 10 years after production of a particular model has ceased. These documents might be the only evidence available to demonstrate that a product was built safely. Records could be kept in either electronic format or hard copy, and it is advisable to keep two copies in separate locations (in case of damage due to fire or other causes).

Documents that should be kept include:

- records of routine test procedures and results
- operation standards related to Safety Critical operations
- Safety Critical component and assembly lists
- final inspection test procedures and results[1]
- records of engineering changes
- calibration records[2].

14.10 Audits

A production control system will only remain effective if it is regularly audited and maintained. A person responsible for its maintenance should be identified (probably from within the quality control or approvals department), and trained to carry out internal audits and procedural

---

[1] For every product tested at end of line, the tested product's serial number, model name/number and test date should be recorded and kept.
[2] A clear policy regarding the calibration of all equipment related to safety testing will be necessary. It should include calibration frequencies (for internal or external calibrations), confirmation procedures, etc. All calibration procedures and records of final safety test equipment including any external calibration certificates should also be kept.

confirmations. A simple check sheet listing areas where safety operations are performed could be prepared and used (see Table 14.5).

Table 14.5 Example of internal audit checklist

INTERNAL AUDIT CHECKLIST

Audit date:	Auditor:	Signed:
AREA	COMMENTS	Pass/Fail
1.0 Incoming goods		
1.1 C of C control*	No problems found	Pass
1.2 Storage conditions	High bay parts areas not marked as safety critical	Fail
etc.		
etc.		
etc.		
2.0 Assembly lines		
2.1 Correct operation standards		
etc.		
etc.		
3.0 Quality control		

*Certificate of Conformance

14.11 Summary

By following the suggested control, inspection and testing procedures as explained above, most manufacturers will be in a good position to demonstrate 'due diligence'.

The necessary level of control will vary between manufacturers, however, depending on the scale of production, the size of the plant, the product range, the complexity of operation and innovation employed. The reader will need to decide on the most appropriate system for his

Chapter 15

Factory control guidelines

If a factory is to produce a safe product at all times, certain guidelines should be set up and followed regarding the handling and control of components and assemblies which have safety implications. When certification is sought from a National Certification Body and the product (thereafter) will have an approval mark affixed, the NCB will require that the manufacturing base (factory) be subject to a programme of regular inspection – usually in the form of an annual audit.

During such a safety audit, the inspector will assess the established systems and procedures so as to give the NCB confidence that (after the initial product testing and approval) *continuing* conformance to the relevant safety standards is being achieved through the manufacturing and testing stages.

In the UK, BEAB issues a document (Document 40) which gives some very useful guidelines on how a manufacturing base should operate. A similar document has recently been issued by CENELEC, CCA-201.

A brief summary of factory control guidelines which may be used is given below.

15.1 Incoming goods area

As a first step, a list of all Safety Critical (SC) components should be prepared by the Design or the Quality Control (QC) departments, and issued to all relevant sections of the plant. In particular, the incoming goods area should use this list to identify all delivered SC components, and thereby accept only approved parts.

Ideally, all incoming Safety Critical components and parts should be accompanied by a valid Certificate of Conformance (C of C) supplied either with every delivery or (through arrangement) periodically – e.g. every 3 or 6 months.

The C of C should clearly identify the part, and indicate all applicable standards or regulations – it must be signed and dated.

When SC parts are purchased in small or irregular lots, then a C of C may not be supplied. In this case the manufacturer has two options:

- to perform 'in-house' testing/confirmation of the components on a sampling basis or
- to create a file containing master copies of C of Cs or other relevant documentation (e.g. copies of technical specifications indicating the components' approval status). This must be updated regularly and used to confirm the validity of incoming components.

Subcontracted SC work need also be accompanied by a C of C confirming that the manufacture of parts, components and/or sub-assemblies has been controlled, and that the relevant guidelines have been followed.

Important – the finished product manufacturer is deemed responsible for all subcontracted work; regular audits of all subcontractors should therefore be performed on a regular basis.

A flowchart is given in Appendix 17 showing how Safety Critical components should be handled throughout the manufacturing process.

15.2 Stores

- All storage areas of SC parts need to be clearly marked and identified as 'Safety Critical' or similar.
- All packages or cartons of SC parts should be stamped (or otherwise marked) as 'Safety Critical' immediately upon receipt (before being placed in their respective storage areas).
- Parts issued from stores to the assembly lines should be confirmed as being of the correct type before issue.
- Lists of SC components should be available in the stores area.

15.3 Assembly lines

- Worksheets including safety assembly operations or the placement of SC parts should be identified as 'Safety Critical' or similar.
- The worksheets issued to assembly operators should clearly indicate the specific SC components or SC assembly operations (of parts/wires, etc.) – e.g. by highlighting these elements in red lettering.
- Drawings or photographs should be provided wherever possible to remove any ambiguity in worksheet instructions relating to Safety Critical placements or assemblies.
- All worksheets should be signed by a responsible person and dated.
- Wherever possible, SC parts should be kept in their original containers.

- Safety Critical parts/assemblies which are classed as 'Reject' or 'Awaiting Repair', etc. should be clearly identified as SC and segregated from non-SC reject parts.
- Wherever possible, the fitting of fuses of similar appearance (but of different value) should be physically separated on the production line to eliminate the possibility of mixing.
- Education of the assembly operators on the importance of safety and the use of good assembly practices is essential – in particular, SC concepts and precautions should be understood.

15.4 End-of-line tests (final safety)

- At the final safety (FS) test position, where *insulation resistance, flash, earth continuity and leakage current* tests are performed, clear procedures and instructions are necessary to explain:
  - how the tests are carried out
  - what equipment to use
  - what action to take in case of failure (i.e. mark with 'REJECT' stamp, etc.)
  - how to investigate and document failures.
- All FS test failures must be recorded and investigated.
- FS tests should be of an appropriate duration (i.e. allowing sufficient test time), and of an acceptable method.
- FS test equipment must be confirmed regularly for correct operation (daily in cases of high volume production) – this may be done using standard test boxes. The confirmation test results need to be recorded, together with the signature of the person authorized to carry out the tests. A written procedure explaining the method of the confirmation test should be kept at the FS test equipment.
- The FS jigs (and any test boxes) must be regularly calibrated.
- External calibrations should take place annually (e.g. HV meter, current meter, DVM, etc.).
- Stickers showing jig calibration expiry dates, last calibration dates, 'Do Not Use After' dates, etc. must be placed on test jigs and boxes.

A flowchart showing safety end-of-line testing and product handling is given in Appendix 17.

All documents, records, calibration procedures, worksheets, failure investigation reports, etc. should be kept for a minimum period of 10 years.

It is very important that safety operations be easily identified. For example, flowcharts such as the example given in Appendix 14 can be used to demonstrate all safety related activities 'at a glance'.

A well-documented factory control procedure would certainly help to show 'due diligence' and thus provide defence in the case of litigation.

Note: *The extent to which the guidelines described in this chapter are implemented will depend on the size of the operation and the volume of production. Accordingly, individual guidelines may be either fully or partially implemented.*

Chapter 16

Quality processes and the LVD

Manufacturers subjected to external audits by a National Certification Body can expect to be audited not only on safety issues, but also on the Quality Control Systems utilized in their operations – regardless of whether or not they be Safety Critical. Application forms which need to be completed before a pre-licence audit by an NCB takes place will require details and evidence of any externally accredited Quality System in place, e.g. ISO 9002, ISO 14001, etc.

During an audit, the visiting inspector will not only access the control of the Safety Critical material and the relevant documentation, but also the Quality Assurance procedures in place and their regular review, in order to guarantee continuing conformance with the procedures as specified in CCA 201 or BEAB Document 40.

16.1 Safety audits

External audits for safety, could be categorized into three types:

- *Inspections* – during these visits the auditor will inspect the Safety Critical materials used, the test methods employed and the results obtained. He will check the documentation, and possibly the actual location and storage of parts.
- *Procedure check* – during these visits the inspector tends to concentrate on an assessment of the manufacturer's procedures and systems in place. He generally spends less time carrying out physical inspections in favour of evaluating the systems and procedures in place, and the evaluation of operating manuals related to Safety Critical and Quality Assurance operations.
- *Product check* – this is a typical Underwriters Laboratory (UL), US

style audit. The inspector will perform a very detailed inspection based mainly on the product rather than the manufacturing systems. A comparison is made between the components, assemblies and materials listed in the report generated (specified) by UL, and those built into the actual manufactured product. Action taken will depend on the severity of any deviations found.

The guidelines issued by CCA (Document CCA 201) are very general and it is up to the manufacturer to decide the scope and nature of the checks undertaken by the site's Quality Assurance function and any other appropriate department, including the extent of internal quality audits. The visiting NCB would like to see in existence a planned schedule of quality audits, and documented evidence demonstrating adherence to the plan.

16.2 Quality system

Having a product tested (internally or externally), having a Test Report in your possession, a Technical File compiled and an EC Declaration of Conformity do not guarantee continued product performance. Procedures for strict manufacturing controls based on a recognized Quality Assurance system, such as ISO 9000, will certainly be helpful to achieve consistency in production.

Meeting the requirements of the LVD in manufacturing operations is, in some respects, no different from meeting any other conformance requirement such as ISO 9000. A Quality System may be set up which can be as complicated or as simple as the manufacturer wishes it to be – the same applies in meeting the requirements of the LVD. ISO 9000 quality procedures can be adopted to include the safety elements required by the LVD, but this can place a large burden on hard-pressed manufacturers who need to produce more goods, of higher quality, faster, and at lower cost. A balance between the cost of setting up such a system and the risk of safety defects slipping through into the market has to be found.

16.3 Procurement of Safety Critical materials

It is essential that the Procurement Officers have a clear understanding of components and materials with Safety Critical implications in the finished product. Safety Critical components lists should be distributed to this section, and then analysed and digested by every member involved in the purchase of such parts. Although Procurement Officers are not usually engineers, it would be very useful to have a reasonable understanding of the usage of such components, and the implications if they fail to operate safely.

Disciplined procedures, preferably under the ISO 9000 regime, will ensure that adequate control is exercised for:

- the purchase of approved (e.g. by the designer) Safety Critical components
- the purchase from approved suppliers only
- the purchase and use of approved alternative parts (in case of supplier change, unavailability of part, etc.).

In some organizations, most functions related to the purchase, inspection and use of Safety Critical parts are performed by the same department or even the same person – in this case it is much easier and simpler to exercise adequate control.

In other (often larger) organizations, there could be a number of departments involved, all performing a different function but all related to each other, for example:

- Design section (designs product with 'safety in mind') – decides which parts perform a Safety Critical function.
- Parts section – prepares specifications for parts requested by the designers, and evaluates suppliers and parts.
- Purchasing section – follows requests from Parts/Design sections for the purchase of materials, negotiates price, and seeks alternative parts/ suppliers if deemed necessary.
- Incoming inspection section – performs sample incoming components checks, monitors and controls supplier approval certificates and other relevant documentation for Safety Critical components. In addition, an advanced vendor management section may operate a supplier audit system and investigate actual parts failures. All of these sections will therefore need to understand each other's functions, have regular meetings and have procedures in place clearly defining their responsibilities. The Procurement department in most organizations operates under extreme pressure – time and cost are the enemy. Searching for an alternative component vendor who can deliver with a lower lead time and at lower cost can often be too tempting – sometimes resulting in the purchase of non-approved or even non-suitable parts!

In addition to 'in-house' understanding and co-operation, we must not forget the vendor and the information either supplied to him or expected from him. For example:

- the supplier must be fully aware that the part he supplies is used in a Safety Critical context
- the supplier may be requested to supply Certificates of Conformance, Approval Certificates, test data or other related documents
- the contract may need to include specific clauses related to the authorization of changes made by the supplier (e.g. in his raw materials

or production processes), or to the original specification of the manufacturer.

From the brief information given above, it should be clear by now that a Quality System should be in place ensuring adequate control of functions and operations involved in the manufacturing of a good quality, and (above all) a safe product.

16.4 Manufacturing control

Having placed adequate quality control procedures in Procurement and Incoming Goods areas, we now have the confidence to concentrate our attention to the task of making a consistent product or assembly.

When a product is submitted for external testing, the manufacturer will usually be required to submit a sample for evaluation, and certification is then awarded based on the construction, components utilized, and wire routing as seen on the submitted sample. Subsequent products should then be produced in accordance with the submitted sample without major deviations. It is impractical of course to assume that during the product's lifetime, changes will not be necessary. Changes of parts suppliers, quality and productivity improvements, etc. will always be necessary. Such changes therefore need to be controlled and approved by a responsible engineer or another qualified person.

Quality control procedures in the manufacturing areas are therefore necessary, clearly indicating operations and procedures essential for the control of changes to (in particular):

- Safety Critical parts
- Safety Critical operations and assemblies
- safety related testing methods
- the method of preparing and approving operation standards
- quality related measurements and testing
- production control systems and the frequency of their application, etc.

Incorporating the safety related aspects of a manufacturing operation into its Quality Manual is far better and more beneficial than the issue of separate documentation.

ISO 9000 can therefore provide the general direction for the issue, authorization and periodical review of assembly work and instructions.

The importance of having a quality process which is designed to additionally identify where the assembly method is critical to product safety cannot be emphasized enough.

The 'queueing' time in safety test laboratories could vary from a few weeks to a few months, and in most cases a manufacturer will submit samples for testing or evaluation from the prototype stage – approval will then be based upon the evaluation of this sample. Subsequent samples through the product's life from early trials through mass production must be confirmed as having been built identically to the originally submitted test sample – this can be achieved by following a documented quality procedure indicating:

- use of the Certification sample returned from the safety laboratory – creation of lists of Safety Critical assemblies and parts can be made using this sample
- evaluation of subsequent trial run samples
- evaluation of a first mass production run sample
- periodic off-line tests – daily, weekly, etc.

The aim being to achieve repeatable and safe product build, based on the construction of the Certified sample.

All key functions of the manufacturing stage need to be documented and reviewed, and without quality procedures in place due diligence will be very hard to prove.

16.5 Quality Assurance functions

In most organizations operating under the ISO 9000 umbrella, the Quality Assurance department is the instigator of such a system and is also responsible for its policing. Quality Assurance procedures should cover:

- quality tests – on and off the production line
- process control checks
- safety tests and evaluations – on and off the line
- calibration tests of quality and possibly safety test equipment
- the control and review of the ISO 9000 procedures, not only for the Quality Assurance section itself, but also (probably) for the whole manufacturing plant
- the introduction of test routines related to product confirmations before mass production begins (i.e. design review)
- the introduction of procedures for analysis of feedback from the market (i.e. analysis of market failures with consequent action for continuous reliability improvement).

16.6 Due diligence procedure

The Low Voltage Directive can have an impact on most functions of an organization involved in the manufacture of electrical products – design and production are not the only sections affected.

It can be very easy to create a Quality Procedure describing actions to be taken by all departments involved in meeting the LVD – thus ensuring and demonstrating due diligence. The procedure itself may be designed so as to include:

- *Purpose* – a clear indication that the manufacturer operates a demonstrable system to ensure compliance with relevant approvals, etc.
- *Scope* – the listing of departments under the scope of this procedure, and their function.
- *Policy* – a description of the company's essential philosophy behind the procedure, and its commitment to take all reasonable precautions necessary to demonstrate due diligence.
- *References* – listing of any other related manuals and procedures.
- *Responsibilities* – manager(s) responsible for establishing, updating and maintaining the procedure and for records retention.
- *Description* – reasons for preparing the procedure and activities carried out by various sections in order to:
  - ensure the production of a safe product
  - ensure customer safety
  - demonstrate the company's activities to any external enforcement agency
  - provide defence in cases of Product Liability claims
  - demonstrate due diligence.

Any systems employed internally to monitor corrective actions, follow up changes, introduce new standards, etc. should also be documented and described in the due diligence procedure.

16.7 Rework/Market recall

Considerations should be given to those procedures associated with the in-house rework of defective products, and also to the recall and rework (or destruction if necessary) of serious market defects.

Procedures should be prepared and listed in the ISO 9000 Quality System describing how a manufacturer can cope with such situations and detailing the actions which need to be taken so as to ensure swift and effective rectification – but in as 'painless a manner as possible'.

In case of in-house rework, points to be considered are:

- location for the rework
- availability of (suitably skilled) manpower
- timing
- purchase of materials which may be necessary in cases involving parts or packaging replacement
- cost implications
- speed of rework
- market opportunity loss
- meeting customer delivery expectations.

In case of market recall, in addition to the above, other items to be considered are:

- tracing the products
- media announcement
- availability of market resources to support the rework as an alternative to returning the product to the factory, e.g. service centres, independent dealers, etc.
- customer dissatisfaction
- cost implications
- loss of customer confidence
- loss of image (reputation) – swift and positive action with good customer.

Communications, however, can actually result in the enhancement of reputation once the initial disruption has been settled.

16.8 Summary

Setting up and maintaining a comprehensive and effective Quality System which incorporates the safety control aspects of a manufacturer's operations is very important. It doesn't necessarily have to be an ISO 9000 accredited system but it could very well be based on the ISO requirements. Manufacturers operating such systems can achieve significant profit increases and better brand image through consequent reductions in product defects (internally and externally) and increased product reliability.

In summary, a good quality system can help a manufacturer to:

- reduce Certification costs (many NCBs recognize an ISO 9000 certification system – resulting in reduced audits and sample requests)
- reduce market defect rates by operating adequate in-house process control and test systems
- increase product reliability by using certified components
- avoid expensive market recalls
- avoid in-house reworks
- meet the LVD – more simply
- produce a better quality and more reliable product.

Appendices

Appendix 1   List of European Union Directives
Appendix 2   List of Harmonized safety standards
Appendix 3   UK Notified Bodies under the Low Voltage Directive
Appendix 4   Other European Certification Bodies/Testing Laboratories
Appendix 5   List of test equipment for electrical safety testing
Appendix 6   Example of the contents of a Harmonized safety standard
Appendix 7   Insulation requirements between parts – guidance examples
Appendix 8   Insulation types, electrical connections and examples
Appendix 9   Creepage distances and clearances – measurement guide
Appendix 10 Test circuit for measuring 'touch currents'
Appendix 11 Test instruments
Appendix 12 Graphic symbols
Appendix 13 Examples of an EC Declaration of Conformity
Appendix 14 Procedure for handling Safety Critical components and operations
Appendix 15 Engineering Change Note – example and completed document
Appendix 16 Production Change Request – example
Appendix 17 End-of-line (final safety) tests
Appendix 18 Useful addresses
Appendix 19 Glossary of terms
Appendix 20 Templates

Appendix 1

List of European Union Directives

Products covered by European Product Safety legislation

The product groups covered by the legislation listed below are all required to carry a CE Mark. The CE Mark is a visible declaration or statement by a manufacturer that his product meets all the requirements of any relevant European legislation. This includes not only the safety elements of the law, but also any requirements to create and keep associated documentation.

Legislation and scope

The Simple Pressure Vessels Directive 87/404/EEC (as amended)
e.g. Paint spraying equipment

The Construction Products Directive 89/106/EEC (as amended)
e.g. Chimney flues, Extractor fans

The Personal Protective Equipment Directive 89/686/EEC (as amended)
e.g. Respirators, Safety glasses

The Telecommunication Terminal Equipment Directive 91/263/EEC (as amended)
e.g. Telephones, Fax machines

The Active Implantable Medical Devices Directive 90/385/EEC (as amended)
e.g. Pacemakers

The Electromagnetic Compatibility Directive 89/336/EEC (as amended)
e.g. Computers, Power tools

The Machinery Directive 89/392/EEC (as amended)
e.g. Circular saws, Petrol lawnmowers

The Hot Water Boilers Directive 92/42/EEC
e.g. Central heating systems

The Medical Devices Directive 93/42/EEC
e.g. Walking sticks, Heart monitors

The Low Voltage Directive 73/23/EEC
e.g. TVs, Toasters

The Toy Safety Directive 88/378/EEC
e.g. Train sets, Teddy bears

The Non-Automatic Weighing Instruments Directive 90/384/EEC
e.g. Weighing scales

The Gas Appliances Directive 90/396/EEC
e.g. Gas fires, Gas barbecues

The Recreational Craft Directive 94/25/EEC
e.g. Boats

The Explosive Atmospheres Directive 94/9/EEC
e.g. Explosimeters

Appendix 2

List of Harmonized safety standards

The list of Harmonized safety standards for the Low Voltage Directive is very extensive (there are a few hundred of these documents already published) and many are supplemented by one or more amendments. As more and more updated versions are being published frequently, the reader is advised to search carefully for the latest version applicable to his product and pay particular attention to the implementation date. It is important to avoid designing or testing a product to a standard that may already be obsolete or about to become obsolete. A small selection of commonly used Harmonized safety standards, giving their reference number and a simplified description of their contents (the descriptions given do not necessarily reflect the exact title of the standard), has been compiled and is given below.

The following are extracts from *Approval* an M&M Business Communications publication.

Components – Miniature fuses

EN 60127-1	General requirements
EN 60127-2	Cartridge fuses
EN 60127-3	Sub-miniature fuses
EN 60127-6	Cartridge fuse holders

Components – Low voltage fuse

EN 60269-1	General requirements
EN 60269-2	Fuses for industrial applications
EN 60269-3	Fuses for household and similar applications

Components – Circuit breakers

EN 60898	Overcurrent protection for household and similar applications
EN 60934	Circuit breakers for equipment
EN 61008-1	Residual current operated circuit breakers (RCCB) without overcurrent protection
EN 61008-2-1	RCCB functionality
EN 61009-1	Residual current operated circuit breakers (RCBO) with overcurrent protection
EN 61009-2-1	RCBO functionality

Components – Industrial plugs and sockets

EN 60309-1	General requirements
EN 60309-2	Contact dimensions

Components – Capacitors

EN 60143	Series capacitors for power systems
EN 60252	AC motor capacitors
EN 60931-1	Shunt power capacitors
EN 60931-2	Shunt power capacitors (ageing and distraction tests)

Components – Automatic electric controls for household and similar use

EN 60730-2-1	Requirements for electrical controls for household appliances
EN 60730-2-2	Thermal motor protectors
EN 60730-2-3	Thermal protectors for fluorescent tube ballasts
EN 60730-2-4	Thermal protectors for hermetic motor-compressors
EN 60730-2-5	Automatic electric burner controls
EN 60730-2-6	Pressure sensing controls
EN 60730-2-7	Timers and time switches
EN 60730-2-8	Water valves
EN 60730-2-9	Temperature sensing controls
EN 60730-2-10	Motor starting relays
EN 60730-2-11	Energy regulators
EN 60730-2-12	Electric door locks
EN 60730-2-15	Boiler water level sensors

Components – Switches

EN 61058-2-1	Cord switches
EN 61058-2-5	Change over selectors
EN 61095	Electromechanical contactors

Components – Other

EN 61204	Power supplies – performance and safety
EN 60742	Isolating transformers
EN 61050	Neon transformers
EN 61131-2	Programmable controllers

Household equipment

EN 60335-1	General requirements
EN 60335-2-2	Vacuum cleaners
EN 60335-2-3	Electric irons
EN 60335-2-4	Spin extractors
EN 60335-2-5	Dishwashers
EN 60335-2-6	Cookers
EN 60335-2-7	Washing machines
EN 60335-2-8	Shavers and hair clippers
EN 60335-2-9	Toasters, grills and similar
EN 60335-2-10	Floor treatment machines
EN 60335-2-11	Tumble dryers
EN 60335-2-12	Warming plates and similar
EN 60335-2-13	Frying pans and deep fat fryers
EN 60335-2-14	Electric kitchen machines
EN 60335-2-15	Appliances for heating liquids
EN 60335-2-16	Waste disposers
EN 60335-2-19	Battery power shavers and hair clippers
EN 60335-2-20	Battery powered toothbrushes
EN 60335-2-21	Storage water heaters
EN 60335-2-23	Appliances for skin or hair care
EN 60335-2-24	Refrigerators and freezers
EN 60335-2-25	Microwave ovens
EN 60335-2-26	Clocks
EN 60335-2-27	UV and IR skin treatment
EN 60335-2-28	Sewing machines
EN 60335-2-29	Battery chargers

and many more.

Household equipment – Automatic electric controls

EN 60730-1	General requirements
EN 60730-2-1	Requirements for electrical control for household appliances
EN 60730-2-2	Thermal motor protectors
EN 60730-2-3	Thermal protectors for ballasts for tubular fluorescent lamps
EN 60730-2-4	Thermal protectors for heretic motor-protectors
EN 60730-2-5	Automatic electric burner controls
EN 60730-2-6	Pressure sensing controls
EN 60730-2-7	Timers and time switches
EN 60730-2-8	Water valves
EN 60730-2-9	Temperature sensing controls
EN 60730-2-10	Motor starting relays
EN 60730-2-11	Energy regulators
EN 60730-2-12	Electric door locks
EN 60730-2-15	Boiler water level sensors
EN 60967	Electric blankets

and many more.

Information Technology Equipment (ITE)

EN 50091-1	Uninterruptible power supplies (UPS)
EN 50098-1	Cable for ISDN basic access
EN 60825-2	Fibre optic communications systems
EN 60950	Safety for ITE including business equipment

Installations – Conduit systems

EN 50086-1	General requirements
EN 50086-2-1	Rigid conduit
EN 50086-2-2	Pliable conduit
EN 50086-2-3	Flexible conduit
EN 50086-2-4	Conduit buried underground

Installations – Low voltage switch gear and control gear

EN 60439-1	Type tested and partially tested assemblies
EN 60439-2	Busbar trunking systems
EN 60439-3	Distribution boards
EN 60439-4	Assemblies for construction sites
EN 60947-1	General rules

152 Electrical Product Safety

EN 60947-2   Circuit breakers
EN 60947-3   Switches, disconnectors and fuse combination units
EN 60947-4-1   Electromechanical contactors and motor starters
EN 60947-5-1   Electromechanical control circuit devices

Instrumentation – Direct acting analogue electrical measuring instruments

EN 60051-1   General requirements
EN 60051-2   Ammeters and voltmeters
EN 60051-3   Wattmeters and varmeters
EN 60051-4   Frequency meters
EN 60051-5   Phase meters, power factor meters and synchroscopes
EN 60051-6   Impedance meters
EN 60051-7   Multi-function instruments
EN 60051-8   Accessories
EN 60051-9   Test methods

Instrumentation – Audio meters

EN 60645-1   Pure tone meters
EN 60645-3   Auditory test signals
EN 60645-4   High frequency audio meters
EN 60651   Sound level meters

Lasers

EN 60825-1   Equipment classification and requirements
EN 60825-2   Fibre optic communications systems
HD 194 S1   Electrical safety

Lighting – Lamp caps and holders

EN 60061-1   Lamp caps
EN 60061-2   Lamp holders
EN 60061-3   Gauges
EN 60061-4   General information

Lighting – Fluorescent lamps

EN 60081   Tubular fluorescent lamps
EN 60155   Glow starters
EN 60400   Lamp holders and starter holders

EN 60901	Single capped fluorescent lamps
EN 61195	Double capped fluorescent lamps
EN 61199	Single capped fluorescent lamps
EN 60920	Ballast for tubular fluorescent lamps
EN 60924	DC electronic ballasts
EN 60926	Starting devices (not glow starters)
EN 60928	AC electronic ballasts

Lighting – Other

EN 60357	Tungsten halogen lamps (not vehicles)
EN 60188	High pressure mercury vapour lamps
EN 60238	Edison screw lamp holders
EN 60432-1	Tungsten filament lamps
EN 60432-2	Tungsten halogen lamps

Lighting – Luminaires

EN 60598-1	General requirements
EN 60598-2-1	Fixed general purpose luminaires
EN 60598-2-2	Recessed luminaires
EN 60598-2-3	Street lighting
EN 60598-2-4	Portable general purpose luminaires
EN 60598-2-5	Floodlights
EN 60598-2-6	Filament lamps and built-in transformers
EN 60598-2-7	Portable luminaires for garden use
EN 60598-2-8	Handlamps

and many more

Machines – Rotating electrical machines

EN 60034-1	Rating and performance
EN 60034-4	Synchronous machines
EN 60034-5	Enclosure protection
EN 60034-6	Methods of cooling
EN 60034-7	Construction and mounting
EN 60034-9	Noise limits
EN 60034-18-1	Insulation systems
EN 60034-18-31	Thermal evaluation of windings

Machines – Low voltage switch gear and control gear

EN 60439-1	Type tested and partially tested assemblies
EN 60439-2	Busbar trunking systems
EN 60439-3	Distribution boards
EN 60439-4	Assemblies for construction sites
EN 60947-1	General rules
EN 60947-2	Circuit breakers
EN 60947-3	Switch disconnectors and fuse units
EN 60947-4-1	Electromechanical contactors and motor starters
EN 60947-5-1	Electromechanical control circuit devices
EN 60947-6-1	Automatic transfer switching equipment
EN 60947-6-2	Control and protective switching devices
EN 60947-7-1	Terminal block and copper conductors
HD 419.2 S1	Semiconductor contactors

Telecommunications

EN 41003	Safety of equipment connected to the network
EN 50098-1	Cable for ISDN basic access
EN 60825-2	Fibre optic communications systems

Television and Radio – Cable distribution for TV and sound

EN 50083-1	General safety requirements
EN 50083-3	Active coaxial wideband distribution equipment
EN 50083-4	Passive coaxial wideband distribution equipment
EN 50083-5	Headend equipment
EN 50083-6	Optical equipment

Television and Radio – Other

EN 60065	Mains operated electronic equipment for household and similar general use
EN 60215	Radio transmitting equipment

Tools – Hand held electric tools

EN 50144-1	General requirements
EN 50144-2-1	Drills
EN 50144-2-2	Screwdrivers and impact wrenches
EN 50144-2-4	Sanders
HD 400.1 S1	General

HD 400.2A S1   Drills
HD 400.2B S1   Screwdrivers and impact wrenches
HD 400.2C S1   Grinders, polishers and disc sanders
HD 400.2D S1   Sanders
HD 400.2E S1   Circular saws and circular knives
HD 400.2F S1   Hammers
HD 400.2G S1   Spray guns
HD 400.3H S1   Metal sheers and nibblers
HD 400.3I S1   Tappers
HD 400.3J S1   Jig saws
HD 400.3K S1   Concrete vibrators
HD 400.3L S2   Chain saws
HD 400.3M S2   Planners
HD 400.3N S2   Hedge trimmers and grass shears
HD 400.3O S1   Routers
HD 400.3R S1   Trimmers

Welding – Arc welding

EN 50060   Power sources
EN 50063   Resistance welding
EN 50078   Torches and guns
EN 60974-1   Power sources
HD 22.6 S1   Cables
HD 22.6 S2   Cables
HD 362 S1   Safety rules
HD 407 S1   Safety rules
HD 427 S1   Installation
HD 433 S1   Coupling devices for welding cables

Definitions

EN: European Norme
HD: Harmonized Document

Appendix 3

UK Notified Bodies under the Low Voltage Directive

AMTAC Laboratories Ltd
Norman Road, Broadheath
Altrincham
Cheshire WA14 EP

ASTA Certification Services
ASTA House, Chestnut Field
Rugby
Warwickshire CV21 2TL

British Approvals Board for
Telecommunications (BABT)
Claremont House
34 Molesey Road, Hersham
Walton-on-Thames
Surrey KT12 4RQ

British Approvals Service for
Cables (BASEC)
Silbury Court
360 Silbury Boulevard
Milton Keynes MK9 2AF

British Electrotechnical Approvals
Board (BEAB), 1 Station View
Guildford
Surrey GU1 4JY

British Standards Institution
Mayland Avenue
Hemel Hempstead
Herts HP2 4SQ

BSI Quality Assurance
PO Box 375
Milton Keynes MK14 6LL

Celectica Ltd
Westfields House, West Avenue
Kidsgrove
Stoke-on-Trent
Staffordshire ST7 1TL

ERA Technology Ltd
Cleeve Road, Leatherhead
Surrey KT22 7SA

ITS Cranleigh (UK) Ltd
Manfield Park, Cranleigh
Surrey GU6 8PY

KTL
Saxon Way, Priory Park West
Hull HU13 9PB

NEMKO Ltd
15 Chelsea Field Estate
Western Road
London SW19 2QA

SGS United Kingdom Ltd
Gaw House, Alperton Lane
Wembley
Middlesex HA0 1WU

SGS United Kingdom Ltd
South Industrial Estate, Bowburn
Durham DH6 5AD

Specialised Technology Resources
(UK) Ltd
10 Portman Road, Reading
Berkshire RG30 1EA

Technology International (Europe)
Ltd
41–42 Shivenham Hundred
Business Park, Shivenham
Swindon
Wiltshire SN6 8TZ

The Lighting Association
Stafford Park 7, Telford
Shropshire TF3 3BD

TRL EMC Ltd
Northern Region Laboratory
Moss View
Nipe Lane, Up Holland
West Lancashire

TÜV Product Services Ltd
Segensworth Road
Titchfield, Fareham
Hampshire PO15 5RH

Appendix 4

Other European Certification Bodies/ Testing Laboratories

Austria

Österreichischier Verband für
Elektrotechnik (ÖVE)
Eschenbachgasse 9
A-1010 Wien

Belgium

Comité Electrotechnique Belge
(CEBEC)
Rhoede 125
Av Nouveau
B-1640 Sint-Genesuis-Rode

INIEX
200 Rue du Chera
B-4000 Liège

Denmark

Demko A/S (DEMKO)
Lyskaer 8
Postboks 514
DK-2730 Herlev

Finland

Samkotartastuskeskus (SETI)
Särkiniementi 3
PO Box 21
SF-002 11 Helsinki 21

FIMKO Ltd
PL 30
FIN-00211 Helsinki

France

Union Technique de l'Electricité
(UTE)
4 Place des Vosages
Courbevoie 92400

Laboratoire Central des Industries
Electriques (LCIE)
33 Avenue du General-Leclerc
F-92260 Fontenay-aux-Roses
Cedex

Germany

Deutsche Elektrotechnische
Kommission im DIN und VDE
(DKE)
Stresemannalee 15
D-60596 Frankfurt/Main

Quelle-Institut für Warenprüfung
Wittekindstrasse 26
D-8500 Nürenberg 80

Otto Versand Warenprüfung
Wandsbeker Strasse 3–7
D-22179 Hamburg

ERG Elektrotechnische
Roetzstr 58
Soellingen Pfinztal 76327

Technischer Uberwachungs-Verein
Stuttgart eV
Prüfstelle für Gertesicherheit
Bernhausen
Gottlieb-Daimler-Strasse 7
D-7024 Filderstadt I

Neckerman Versand AG
Warenprüfung
Hanauer Landstrasse 360-400
D 60386 Frankfurt/Main
Hessen

Hellas

ELOT
Didotou 15
GR-106 80, Athens

Ireland

Electro-Technical Council of
Ireland
Institute for Industrial Research
and Standards

Ballymun Road
IRL-Dublin 9

Italy

Istituto italiano del marchio di
qualità (IMQ)
Via Quintiliano 43
1-20138 Milano

Centro Elettronico Sperimentale
Italiano
Giacinto Motta SpA – CSEI
Via Rubattino 54
1-20134 Milano

Norway

Norges Electriske
Materilikontroll – NEMKO
Gaustadalleen 30
N-0314
Oslo

Portugal

Instituto Português da Qualidade
(IPQ)
Rua C a Avednida dos Três Vales
P-2825 Monte de Caparica

Spain

Asociación Española de
Normalización y Certificación
(AENOR)
C/Fernandez de la Hoz, 52
E-28010 Madrid

Asociación Electrotécnica
Española (AEE)
Francisco Gervas 3
E-28020 Madrid

Sweden

Svenska Elektriska Kommissionen (SEK)
PO Box 1284
Kistagången 19
S-164 28 Stockholm

SEMKO AB (SEMKO)
Box 1103
S-164 22 Kista-Stockholm

Switzerland

Schweizerischer Elektrotechnischer Verein (SEV)
Seefeldstrasse 301
Postfach CH-8034 Zurich

The Netherlands

KEMA NV (KEMA)
Utrechtseweg 310
NL-6812 AR
Arnhem
(Postbus 9035 – L-6800 EF Arnhem)

Appendix 5

List of test equipment for electrical safety testing

Instrument	Comments (possible uses)
Digital voltmeter (DVM)	Used for leakage current measurement during short and open circuit tests
Calliper	Used for creepage distance and clearance measurement
Micrometer	Used for insulation thickness measurement
Test pin	Used to assess electric shock hazard – could be made 'in-house' according to a standard's drawings (see also Appendix 11)
Test finger	Used to assess electric shock hazard – could be made 'in-house' according to a standard's drawings (see also Appendix 11)
Taper pin	Used to assess electric shock hazard – could be made 'in-house' according to a standard's drawings (see also Appendix 11)
Impact Test hammer	Used to assess electric shock hazard
Small oven	Used for conditioning before flammability tests – ageing cycle for insulation tests
Environmental chamber	Used for conditioning components before testing – temperature and humidity test

Tester for:
  Dielectric test
  Earth continuity
  Insulation resistance
	Used for insulation stress testing, for confirmation of protective earth and to confirm that insulation resistance meets the requirements

Digital multichannel temperature meter	Used for temperature measurement under normal and fault operating conditions

(*continued*)

Instrument	Comments (possible uses)
Thermocouples	Used for temperature measurements as above – can be calibrated internally against a known externally calibrated thermocouple
Temperature/Humidity meter	Used for calibration check of environmental chamber – and the ongoing monitoring of test conditions
Milliammeter	Used for dielectric strength (flash) test current measurement
Power supply	Used for providing a stable energy source
Power supply (DC)	Used for powering a relay when applying a fault test
Wattmeter	Used for power consumption measurement during normal and standby operating conditions

Appendix 6

Example of the contents of a Harmonized safety standard (summary only)

Extract based on BS EN 60065: 1994 – *not latest issue*

Title

'Safety requirements for mains operated electronic and related apparatus for household and similar general use.'

Scope

Details the type of equipment covered by the standard, e.g. radio receiving apparatus for sound or vision, amplifiers, electronic musical instruments, battery eliminators, etc.

Does not apply to equipment designed for supply above 250 V (rms) or 433 V (rms) between phases in the case of three phase supply.

Clause 2: Definitions

The clause describes definitions for the terms used in various clauses of the standard, e.g. live part denotes a part, contact with which may cause significant electric shock.

Clause 3: General requirements

In this section it is specified that the equipment shall provide protection against:

- electric shock
- excessive temperature
- radiation
- implosion
- fire
- mechanical instability.

Clause 4: General condition for tests

Specifies conditions for test, atmospheric conditions, nature of supply, signal conditions, control settings, fault conditions, etc.

Clause 5: Marking

In this clause the standard specifies that the model name label should be indelible, legible, and placed on the exterior (excluding the base).

The label is tested by rubbing lightly for 15 s with a cloth soaked in water, then (in a different place) with a cloth soaked in petroleum spirit.

The model name label should contain:

- manufacturer's name or trademark.
- model number or model name
- double square symbol (Class II)
- nature of supply
- voltage range 220–240 V
- mains frequency 50 Hz.

User instructions – must contain any relevant safety information, and must be in the official language of the country to which the product is being supplied.

Clause 6: Ionizing radiation (applicable to television products)

For this test, the standard specifies that the controls are adjusted to give maximum radiation while maintaining an intelligible picture.

The measurements are made after 1 hour. Measured levels must be less than 0.5 mR/h measured at 5 cm from the outer surface of the television.

Clause 7: Heating under normal operating conditions

This test is intended to ensure that the equipment does not fail or become unsafe in use due to overheating. The test is carried out in the least favourable way. The Equipment Under Test (EUT) is placed in the normal position of use and ventilation – according to the instructions supplied by the manufacturer.

If the position of use is not specified, then the EUT is positioned 5 cm behind the front edge of a wooden test box with 1 cm free space along the sides and top and 5 cm along the rear side.

The EUT is connected to a supply voltage of 0.9 or 1.06 times the rated voltage or any nominal supply voltage (whatever provides the worst case).

In the case of a television, the test is performed when receiving a normal picture and at 1/8 the rated or non-clipped audio output power, the audio signal being white noise.

Temperature rises are measured when a steady state has been reached – assumed to be 4 hours maximum.

Temperature rise above ambient must not exceed 85°C for Printed Wire Boards (PWBs) and winding wires, and not exceed 60°C for the enclosure.

Clause 8: Heating at elevated temperatures?   Deleted

Clause 9: Shock hazard under normal operating conditions

Accessible parts must not be live, in addition terminal devices must not be live even if inaccessible. Tested using a rigid or jointed test finger – force of 50 N is applied on all outer surfaces of the EUT including the base. Also a tapered test pin is applied to all apertures without appreciable force. Neither test finger nor pin must become live.

To determine whether a part or terminal is live:

- For antenna and earth, the current measured through a non-inductive resistor of $2\,k\Omega$ must be less than 0.7 mA pK.
- For all other parts or terminals the current measured through a non-inductive resistor of $50\,k\Omega$ must be less than 0.7 mA pK.

Ventilation holes

Must be designed such that a suspended foreign body (for example, a necklace) cannot touch live parts. Tested by inserting through the holes a

4 mm × 100 mm test pin, suspended freely from one end – the pin must not become live.

Terminal devices

Making a connection to a terminal device must not cause a shock hazard. The test is performed by using a test pin (1 mm × 20 mm) within 25 mm of every terminal with a force of 10 N.

Also, each terminal is tested with a straight bare wire of 1 mm × 100 mm – the pin or wire must not become live.

Pre-set controls

Adjustment from outside the enclosure using screwdrivers or other tools must not present risk of electric shock. Tested by inserting a metal pin (2 mm × 100 mm) through the access hole – it must not become live.

Withdrawal of mains plug

There must be no risk of shock when touching the pins of the mains plug after it is withdrawn from the socket outlet. A test is performed with the EUT's mains switch in the most unfavourable position – the pins of the plug must not be live 2 seconds after the plug has been withdrawn. The test is repeated up to 10 times.

Resistance to external forces

The enclosure must be resistant to external forces. This is tested by using the rigid test finger with a force of 50 N inwards for 10 seconds and by using the test hook with a force of 20 N outwards for 10 seconds.

Distances between accessible parts and live parts must not be less than those in Table II of the standard and live parts must not become accessible. Class II equipment must be either double insulated (basic plus supplementary insulation), or have reinforced insulation. 6 mm in air is satisfactory provided it is maintained by rigid construction.

Removal of protective covers

Parts that become accessible through removal by hand of a cover must not be live.

Constructional requirements

The insulation of live parts must not be by hygroscopic materials – checked by inspection and the tests of Clause 10.

Product must be constructed so that there is no risk of electric shock from accessible parts.

Class II equipment – accessible parts must be insulated from live parts by either double insulation (basic plus supplementary insulation) or reinforced insulation.

For double insulation either the basic insulation or the supplementary insulation must be 0.4 mm thick. The other insulation can be thinner provided that it withstands the dielectric strength test of Clause 10 for basic or supplementary insulation, i.e. 1.414 kV rms.

Reinforced insulation should have a thickness of 2 mm. Thinner insulation is allowed (0.4 mm) provided it is not subjected to mechanical stress that might lead to deformation or deterioration. It must also withstand the dielectric strength test of Clause 10, i.e. 3 kV rms.

Components meeting the requirements of Clause 14.1 (resistors) or 14.3 (inductors) can bridge basic, supplementary, double or reinforced insulation.

Components meeting the requirements of Clause 14.2.1(b) can bridge double or reinforced insulation (Y1 capacitors). The external insulation of these capacitors must not bridge double or reinforced insulation unless this external insulation meets the requirements of the standard. Generally this insulation is considered unsatisfactory at the wire terminations.

Creepage and clearance distances must be 6 mm (min.) between primary and secondary unless they are maintained by rigid construction (i.e. PWB) when a 1 mm reduction can be applied.

Where they are not maintained by rigid construction the distances are checked using a force of 2 N on all parts. Where relevant, a force of 50 N with the rigid test finger is applied to any point on the outside of the enclosure.

For Class II, double insulation is required between accessible parts and conductors in cables conductively connected to the supply mains (generally this only applies to the mains flexible cord but sometimes to link wires from the input connection to the main PWB). This is more likely to be a problem on VCRs and satellite receivers due to the metal case.

Construction must not allow bridging of insulation due to loosening of screws, etc. This is checked by tests of Clause 12 (bump or vibration test).

The construction must be such that if a wire becomes detached, the creepage distances and clearances are not reduced below those required (i.e. 6 mm between primary and secondary) by the natural movement of the wire.

This requirement is met if there is no possibility of the wire becoming detached. This is covered by wires having a mechanical as well as electrical fixing. Examples of a mechanical fix are wires twisted together, fastened together with tape, having a wrap joint, glued to PWB, etc. It must always be assumed that a solder joint by itself will fail.

Fastening of windows, lenses, etc. must be adequate if live parts are accessible when they are removed. Test is performed by applying an outward force of 20 N for 10 seconds in the most unfavourable position.

Clause 10: Insulation requirements

Insulation must withstand surges:

50 discharges at a maximum of 12 per minute from a 1 nF capacitor charged to 10 kV are applied between aerial and mains terminals. Then the insulation is measured at 500 V DC – must be more than 2 MΩ.

Humidity treatment

The product must not become unsafe due to the humid conditions which may occur in normal use.

The product is tested by placing it in an environmental chamber for 48 hours at 30°C and 90–95% RH (for tropical conditions: 7 days at 40°C and 90–95% RH).

Insulation resistance and dielectric strength

This is measured immediately after completion of the humidity treatment.

Dielectric strength is performed first by applying 3 kV rms for 1 minute:

- across mains poles
- between mains poles and accessible parts or terminals.

No flashover or breakdown should occur.

Insulation resistance is measured next at 500 V DC across the above points and must be:

- above 2 MΩ for mains poles
- above 4 MΩ for mains poles and accessible parts or terminals.

Clause 11: Fault conditions

All safety standards have a requirement that equipment be safe under abnormal operating conditions. Electronic circuits are tested by simulating component failure that could occur during normal use and might cause the equipment to become unsafe. Only one simulated fault or abnormal condition is applied at a time.

Shock hazard

Protection against electric shock must exist under fault conditions. The temperature of parts acting as a support or barrier must not reach unsafe levels (levels likely to cause mechanical failure) such that live parts become accessible, or creepage distances and clearances are significantly reduced.

Heating

Parts must not reach a temperature such that there is a danger from fire or abnormal heat, or that flammable gases are emitted. Generally this affects only PWBs for which the limit is 110°C rise above ambient. This temperature can be increased provided that not more than $2\,cm^2$ of PWB are heated and solder does not become molten.

Clause 12: Mechanical strength

Equipment must be constructed to withstand the handling expected in normal use.

Bump test

The EUT is placed on a horizontal support of wood and allowed to fall 50 times from a height of 5 cm onto a wooden table. It must not become unsafe (no reduction in creepage and clearances, etc.).

Vibration test (portable equipment)

The test is of 30 minutes' duration at an amplitude of 0.35 mm vertically and frequencies of 10 Hz, 55 Hz, and 10 Hz at sweep of approx. 1 octave per minute. The product must not become unsafe due to reduction in creepage distances and clearances (or loosening of fixing screws, etc.).

Impact test

The EUT is subjected to three blows from an impact hammer (0.5 Nm) at every point which provides protection to live parts. This is also applied to windows, lenses, etc. if they protrude from the enclosure more than 5 mm or the area projecting exceeds $1\,cm^2$.

There must be no damage affecting the safety of the product, the enclosure must not have any visible cracks, live parts must not become accessible and insulating barriers must not be damaged.

Clause 13: Parts connected to supply mains

These are parts of the set which are in electrical connection with the supply mains such that a connection between the part and either pole of the supply mains causes in that connection a current greater than 9 A (rupturing current of a 6 A fuse).

Creepage and clearances – must be 3 mm, except that 2 mm is allowed if it is maintained by rigid construction – but only between tracks on a PWB and not pads.

Note: *Creepage distances on PWBs parallel to fuse holders and contacts of switches must meet the requirements for basic insulation (3 mm).*

Clause 14: Components

Resistors

Since the short-circuiting of the barrier resistor would infringe on the requirements for operation under fault conditions, it must meet the creepage and clearance requirements between end cap terminations and be stable under overload. Only approved types should be used.

Capacitors

Capacitors, the short-circuiting of which would infringe the requirements for shock hazard under fault conditions, must be approved to IEC 384-14, 2nd edition – and must be Y1 between live parts and accessible parts.

Capacitors connected between the poles of the supply mains must be approved to IEC 384-14, 2nd edition and must be X2 type (X1 capacitors are for equipment intended for fixed connection to supply mains).

Inductors, Relays, Degaussing coils, Isolation transformers

This subclause also covers constructional requirements for opto isolators and must provide adequate protection against electric shock, i.e. they must meet the requirements of Clause 14.3.2(a) or 14.3.2(b).

14.3.2(a) imposes constructional requirements and a dielectric strength test.
14.3.2(b) has a life test plus constructional requirements.

Isolation transformers

Creepage and clearances must meet requirements for reinforced insulation. The coil former must be at least 0.4 mm thick, and there must be reinforced insulation between primary and secondary windings. The samples must pass 3 kV rms dielectric strength test for 1 minute after humidity treatment of Clause 10.2.

Note: *If Triple Insulated Wire (TIW) is used in the construction then:*

- *the transformer must pass Clauses 14.3.2(a) and 14.3.2(b) or*
- *there must be interleaved insulation if there is mechanical stress on the wires or*
- *the TIW must meet the requirements of Annex U of EN 60950 and subclause 2.9.4.4 of EN 60950 Amendment 4.*

High voltage transformers and components (above 4 kV)

Must be resistant to fire. For high voltage transformers, three samples are pre-conditioned by applying a power of 10 W (at mains frequency) to the high voltage winding for 2 minutes. This is then increased in 10 W steps at 2 minute intervals to 40 W. The treatment lasts 8 minutes or is terminated as soon as interruption of the winding or splitting of the cover occurs.

After being allowed to cool to room temperature the samples are placed in an oven for 2 hours at $100 \pm 2°C$. The samples are then subjected to a flame test using a butane gas flame of 12 ± 2 mm from a burner of diameter 0.5 ± 0.1 mm. The flame is applied for 10 seconds, then 1 minute, and then for 2 minutes. Any self-sustaining flame must go out within 30 seconds.

All other high voltage components are subjected to the flame test only (i.e. HV cables, CRT base sockets).

Fusing devices

Fuse links must comply with IEC 127, fusing characteristics must be marked on the fuse holder or adjacent to it:

e.g. T5AL or T5AH.

Fusing resistors

Must have adequate rupturing capacity. This is normally covered by the fault testing of Clause 11.

Switches

A manually operated mechanical switch is required for equipment which, under normal operating conditions, has a power consumption >15 W and/or a peak voltage exceeding 4 kV. This switch must be connected such that, when it is in the 'off' position the power consumption does not exceed 15 W and peak voltages do not exceed 4 kV even under fault conditions.

The 'on' position of the switch must be clear – this can be achieved by marking, illumination or audible indication. There must be a clearly discernible indication of the standby mode – the switch must be of an approved type.

The switch must be clearly identified – type reference, manufacturer, rated voltage, rated current and rated peak surge current or the ratio between rated current and rated peak surge current – it must also have suitable characteristics for the application.

Batteries

There must be no risk of degradation of insulation. Special requirements exist for Lithium batteries.

Clause 15: Terminal devices

Plugs and sockets

Mains plugs and sockets must be of an approved type. Other connectors – it must not be possible to insert other connectors in a mains socket (e.g. banana plugs).

Protective earth terminal – Class I

Generally covers security of screw terminals and non-consolidation with solder. The standard also has a requirement that conductors of mains flexible cords are not directly soldered to the conductors of the PWB.

Clause 16: Mains cords

Mains flexible cords must comply with IEC 127 or IEC 245. Generally, the cords of TV and video products should comply with IEC 127 – polyvinyl chloride insulated cables.

The mains cord must be of adequate cross-sectional area such that if a short circuit occurs in the receiver, the protection circuits in the electrical installation operate before the cord overheats (0.75 mm^2 will satisfy all national standards).

Connecting points must be relieved from strain and the conductors must be prevented from twisting. Apertures for cords must not cause damage to the cord, and bushings of insulating material must not deteriorate in normal use. This is checked by inspection and by an ageing test for bushings.

Clause 17: Electrical connections and mechanical fixings

Screw fixings which may be loosened and tightened during the life of the equipment must have adequate strength.

This is checked by loosening and tightening 10 times at the specified torque (typically 1.2 Nm for TV backcover screws) – there must be no deterioration which could affect the safety of the product.

There must also be means of ensuring correct location (by a lead to the screw) to prevent fitting in a slanting manner. If the fixing screws are not captive then creepage and clearance distances are checked by using a screw of *length* 10 times the nominal *diameter* of the correct screw (i.e. typically 40 mm) at the torque specified above – there must be no reduction in creepage distances and clearances below those allowed.

Stands or detachable legs supplied by the manufacturer must be supplied with the fixing screws or they must be supplied with the product.

Clause 18: Picture tubes

Must be an approved type. For approval testing, 12 samples are required, half of which are aged in an environmental chamber.

These samples are then subjected to the following tests:

- *Mechanical strength* – impact from hardened steel ball (R62) 40 mm diameter from a height of 210 cm. No glass particles exceeding 10 g should pass a barrier 150 cm from the tube.
- *Implosion* – an area is scratched with a diamond stylus then cooled with liquid nitrogen until a fracture occurs. No glass particles exceeding 2 g should pass a barrier 50 cm from the tube and no particles should pass a barrier 200 cm from the tube.

Clause 19: Mechanical stability

Equipment must be stable in use. This is tested by placing the equipment on a plane inclined at 10° to the horizontal and rotated through 360°. A force of 100 N is directed vertically downward on any (normally) horizontal surface to provide the maximum overturning moment.

Clause 20: Resistance to fire

- *Printed Wire Boards* – PWBs of surface area >25 cm^2 in television receivers must be fire retardant – they should be approved to EN 60065/IEC 65.
- *Backcover* – backcovers and parts of the enclosure having ventilation holes designed for letting out heated air must be of slow burning or fire retardant material. The material should be approved to EN 60065/IEC 65, ensuring that it meets the requirements at the thickness used in the receiver.

Appendix 7

Insulation requirements between parts – guidance examples

(extracts from BS EN 61010-1: 1993)

B: a test for BASIC INSULATION is required
A: a test for DOUBLE or REINFORCED INSULATION is required

HAZARDOUS LIVE means HAZARDOUS LIVE in normal operating conditions

Electrical Product Safety

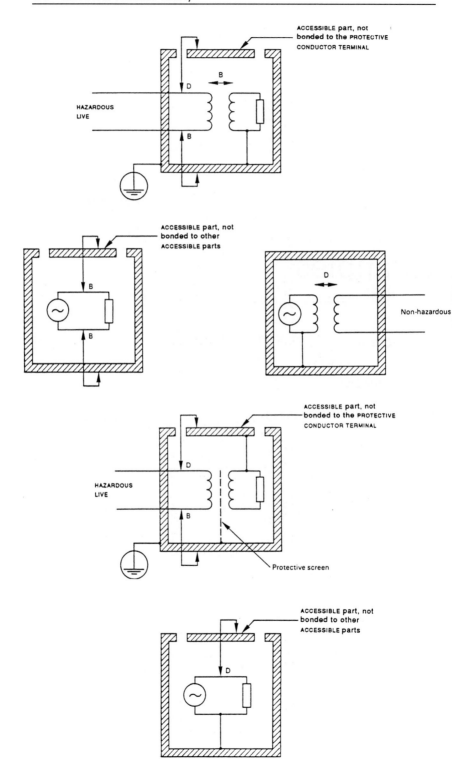

Insulation requirements between parts – guidance examples

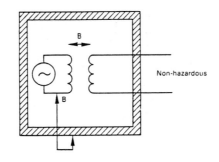
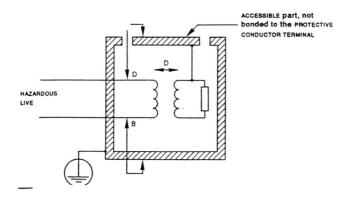
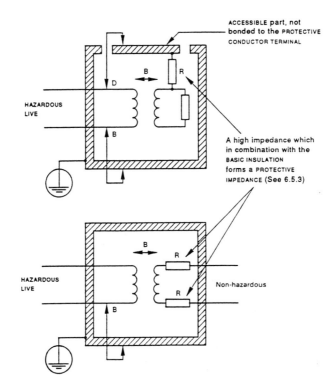

178 Electrical Product Safety

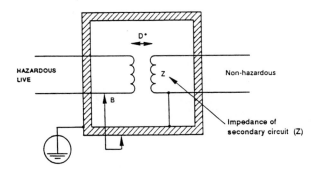

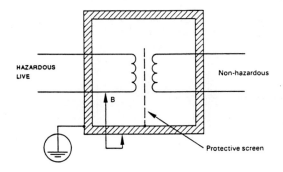

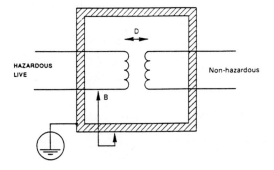

Appendix 8

Insulation types, electrical connections and examples

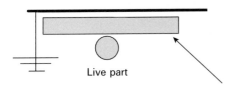

Accessible EARTHED metal part (i.e. metallic enclosure) – **Class I**

Live part

Basic insulator can be used, no minimum thickness, but must withstand a high voltage stress test
or
Minimum distance air gap without the need for an insulator, as long as the live conductor is reliably supported

Figure A8.1 Basic insulation

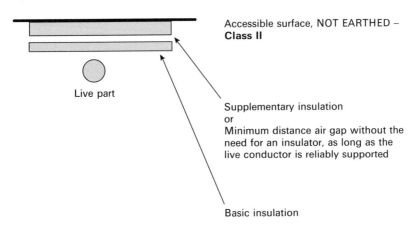

Accessible surface, NOT EARTHED – **Class II**

Live part

Supplementary insulation
or
Minimum distance air gap without the need for an insulator, as long as the live conductor is reliably supported

Basic insulation

Figure A8.2 Supplementary insulation

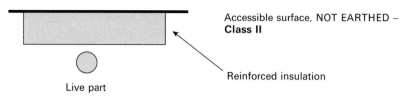

Figure A8.3 Reinforced insulation

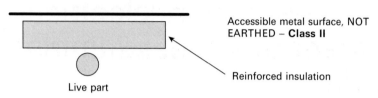

Figure A8.4 Reinforced Insulation

Appendix 9

Creepage distances and clearances – measurement guide

(extracts from BS EN 60439-1: 1994)

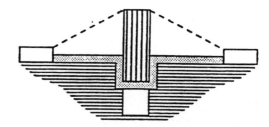

Condition: Creepage distance through uncemented joint is less than creepage distance over barrier.
Rule: Clearance is the shortest direct air path over the top of the barrier.

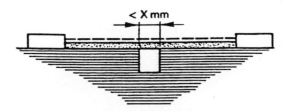

Condition: This creepage distance path includes a parallel- or converging-sided groove of any depth with a width less than X mm.
Rule: Creepage distance and clearance are measured directly across the groove as shown.

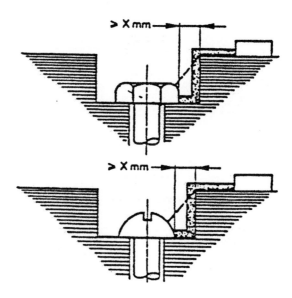

Condition: Gap between head of screw and wall of recess wide enough to be taken into account.
Rule: Clearance and creepage distance paths are as shown.

Creepage distances and clearances – measurement guide

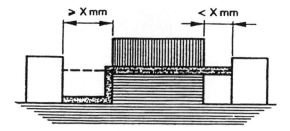

Condition: This creepage distance path includes an uncemented joint with a groove on one side less than X mm wide and the groove on the other side equal to or more than X mm wide.
Rule: Clearance and creepage distance paths are as shown.

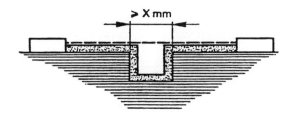

Condition: This creepage distance path includes a parallel-sided groove of any depth and equal to or more than X mm.
Rule: Clearance is the 'line of sight' distance. Creepage distance path follows the contour of the groove.

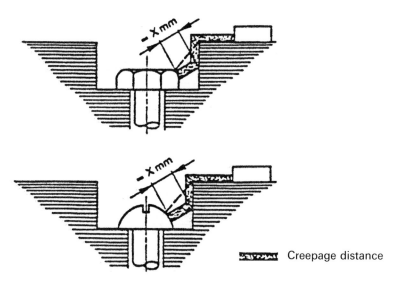

▬▬▬ Creepage distance

Condition: Gap between head of screw and wall of recess too narrow to be taken into account.
Rule: Measurement of creepage distance is from screw to wall when the distance is equal to X m.

184 Electrical Product Safety

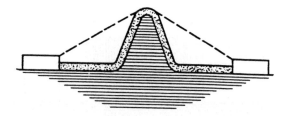

Condition: This creepage distance path includes a rib.
Rule: Clearance is the shortest direct air path over the top of the rib. Creepage path follows the contour of the rib.

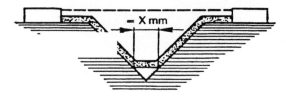

Condition: This creepage distance path includes a V-shaped groove with a width greater than X mm.
Rule: Clearance is the 'line of sight' distance. Creepage distance path follows the contour of the groove but 'short-circuits' the bottom of the groove by X mm link.

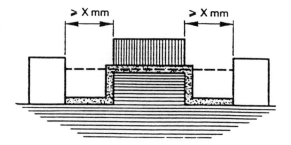

Condition: This creepage distance path includes an uncemented joint with grooves equal to or more than X mm wide on each side.
Rule: Clearance is the 'line of sight' distance. Creepage distance path follows the contour of the grooves.

Creepage distances and clearances – measurement guide 185

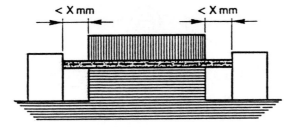

Condition: This creepage distance path includes an uncemented joint with grooves less than X mm wide on each side.
Rule: Creepage distance and clearance path is the 'line of sight' distance shown.

Appendix 10

Test circuits for measuring 'touch currents'

(extract from BS EN 60065: 1998)

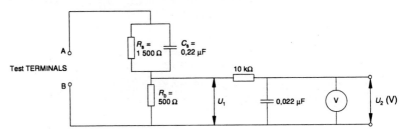

Resistance values in ohms (Ω)

V: Voltmeter or oscilloscope
(r.m.s. or peak reading)

BS EN 60065: 1998 – Annex D (normative) – Measuring network for touch currents according to IEC 60990.

Appendix 11

Test instruments

(extract from BS EN 60065: 1994)

All dimensions are in mm.

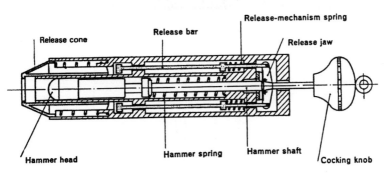

Impact test hammer

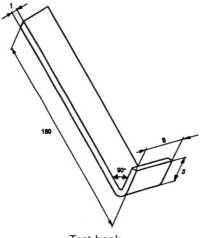

Test hook

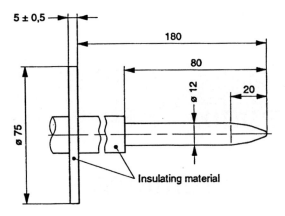

Rigid test finger

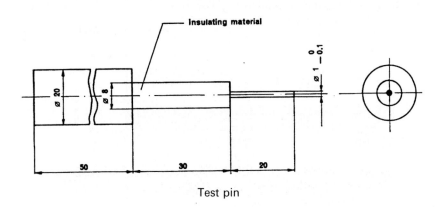

Test pin

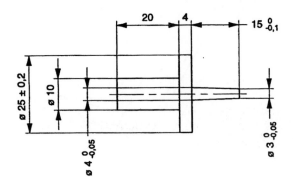

Tapered test pin

Test instruments 189

Impact test hammer

Test hook

Rigid test finger

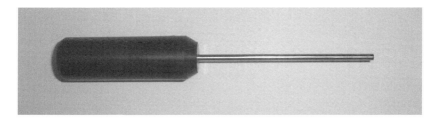

Test pin

Tapered test pin

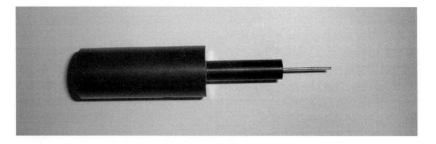

Stepped pin

Appendix 12

Graphic symbols

(extract from BS EN 61010-1: 1993 & A2: 1995)

Number	Symbol	Publication	Description
1	⎓	IEC 417, No. 5031	Direct current
2	∿	IEC 417, No. 5032	Alternating current
3	≂	IEC 417, No. 5033	Both direct and alternating current
4	3∿	IEC 617-2, No. 02-02-06	Three-phase alternating current
5	⏚	IEC 417, No. 5017	Earth (ground) terminal
6	⏚ (circled)	IEC 417, No. 5019	Protective conductor terminal
7	⏚ (frame)	IEC 417, No. 5020	Frame or chassis terminal
8	⏚ (equipotential)	IEC 417, No. 5021	Equipotentiality

(continued)

Number	Symbol	Publication	Description
9	│	IEC 417, No. 5007	On (supply)
10	○	IEC 417, No. 5008	Off (supply)
11	▣	IEC 417, No. 5172	Equipment protected throughout by double insulation or reinforced insulation (equivalent to Class II of IEC 536 – see annex H)
12 (see note)	⚡ (Background colour – yellow; symbol and outline – black)	ISO 3864, No. B.3.6	Caution, risk of electric shock
13 (see note)	♨ (Background colour – yellow; symbol and outline – black)	IEC 417, No. 5041 (417-IEC-5041)	Caution, hot surface
14 (see note)	⚠ (Background colour – yellow; symbol and outline – black)	ISO 3864, No. B.3.1	Caution (refer to accompanying documents)
15	⎴	IEC 417, No. 5268-a (417-IEC-5268-a)	In-position of a bistable push control
16	⎵	IEC 417, No. 5269-a (417-IEC-5269-a)	Out-position of a bistable push control

Note: *Colour requirements for symbols 12, 13 and 14 do not apply to markings on equipment provided that the symbol is moulded or engraved to a depth or raised height of 0.5 mm, or that the symbol and outline are contrasting in colour with the background.*

Appendix 13

Example of an EC Declaration of Conformity

Company name/Address etc.
....
....
....

EC
Declaration of Conformity

We .... under our sole responsibility declare that the product as listed below

Product category: ....

Model name: ....

- conforms with the principal safety objectives of the **European Directive 73/23/EEC**, *[for UK only – as implemented by the Electrical Equipment (Safety) Regulations 1994]*, by application of the following standards: e.g. ***EN 60950: 1992***

    Year of affixation of the CE Marking: *e.g. 1998*

- conforms with the protection requirements of **European Directive 89/336/EEC**, *[for UK only – as implemented by the Electromagnetic Compatibility Regulations 1992]*, by application of the following standards: *e.g. EN 55022: 1995*

Signed:

Title:

Place:

Date:

Company name/Address etc.
....
....
....

EC
Declaration of Conformity

We .... under our sole responsibility declare that the product as listed below complies with the principal safety objectives of the:

Low Voltage Directive 73/23/EEC as amended by 93/68/EEC

Product category: ....

Model name: ....

Applicable standard(s): ....

Year of affixation of the CE Marking: ....

Signed:

Title:

Place:

Date:

Appendix 14

Procedure for handling Safety Critical components and operations

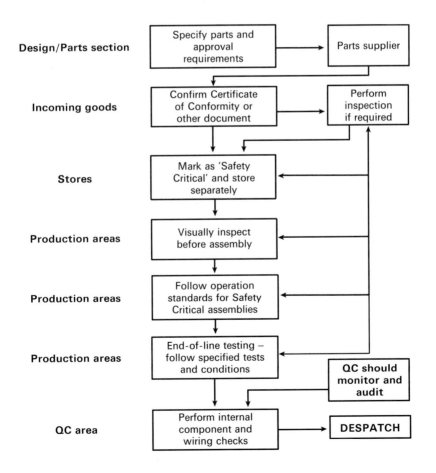

Appendix 15

Engineering Change Note – example and completed document

Example

Company name/logo	Engineering Change Note		Planned by:	Written by:	Checked by:

Subject:		ECN no.:	Issue Date:	Approval: (Design Manager)	Approval: (Purchasing)

Purpose: *e.g. design change, specification change, productivity improvement, production reason, etc.*

Distribution:	Quality assurance	Production eng.	Production	Incoming goods	Accounts	Sub-contractor	Mechanical eng.
	Parts group	etc.					

Model(s) (list model affected)	Suffix (if applicable)	Reference no. (PWB ref. no.)	Old/New part no. (of affected part)	From (value/type)	To (value/type)	Accumulation (start/finish serial no.)	APPROVAL SAFETY
							APPROVAL EMC

Completed document

Engineering Change Note

TP Communications			
		Planned by: K.R. (10.12.98)	Written by: T.N. (11.12.98)	Checked by: R.W. (13.12.98)

Subject: Resistor value changes

ECN no.: R22301 **Issue Date:** 16.12.1998

Approval: (Design Manager)	Approval: (Purchasing)
R.K.J. (18.12.98)	D.M. (19.12.98)

Purpose: design change - EMC performance improvement until new suffix IC TDK 665 is introduced.

Distribution: Quality assurance | Production eng. | Production | Incoming goods | Accounts | Sub-contractor | Mechanical eng.

Model (s) (list model affected)	Parts group	Suffix (if applicable)	Reference no. (PWB ref. no.)	Old/New part no. (of affected part)	From (value/type)	To (value/type)	Accumulation (start/finish serial no.)
e.g. TX-100		-1A	R 101	337-171/336-236	8.2K Ohm/0.1W	5.6K Ohm/0.1W	301-400
TX-100		-1A	R 101	336-236/331-171	5.6M Ohm/0.1W	8.2M Ohm/0.1W	401-

APPROVAL SAFETY	APPROVAL EMC
S.H.D. (17.12.98)	G.M. (17.12.98)

Appendix 16

Production Change Request – example

Production Change Request

Company name/logo			Planned by:	Written by:	Checked by:
Subject:		PCR no.:	Issue Date:		
			Approval: (Design Manager)		**Approval:** (Purchasing other)

Purpose: *e.g. design change, specification change, productivity improvement, production reason, etc.*

Distribution:	Quality assurance	Production eng.	Production	Incoming goods	Accounts	Sub-contractor	Mechanical eng.
	Parts group	etc.					

Model(s)					
			Indicate here (by text and drawing or photograph) the desired manufacturing change	**APPROVAL SAFETY**	
				APPROVAL EMC	

Appendix 17

End-of-line (final safety) tests

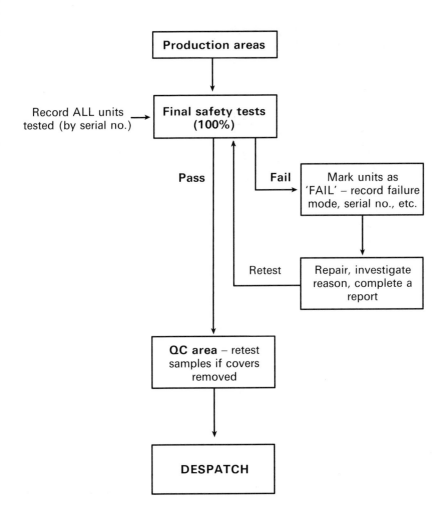

Appendix 18

Useful addresses

The organizations given in Appendices 3 and 4 will provide advice on standards, consultation services and testing facilities. However, if the manufacturer decides to perform his own 'in-house' testing and confirmations, he will require specialized test instruments and equipment as given in Appendices 5 and 11. Addresses of some suppliers that provide test equipment are given below.

Suppliers of specialized test tools and equipment

Testing
Trzaska c. 2
1000 Ljubljana
Slovenia
Tel: +386 (0)61 126 2199

Friborg Instruments AB
PO Box 1, S-738
21 Norberg
Sweden
Tel: +46 (0)223 228 67
Fax: +46 (0)223 228 77

Educated Design and Development, Inc.
521 Uwharrie Ct
Raleigh
NC 27606, USA
Tel: +1 (0)919 821 7088
Fax: +1 (0)919 821 1939

Suppliers of electrical safety test equipment

EMV Ltd
17/18 Drakes Mews
Crownhill
Milton Keynes, Bucks
MK8 0ER, UK
Tel: +44 (0)1908 566556
Fax: +44 (0)1908 560062

Appendix 19

Glossary of terms

A	Amps
AC	Alternating current
ASTA	Association of Short Circuit Testing Authorities
BEAB	British Electrotechnical Approvals Board
BSI	British Standards Institution
BS	British Standards
CCA	CENELEC Certification Agreement
CENELEC	European Electrotechnical Standards Committee
CRT	Cathode ray tube
C of C	Certificate of Conformance
CPA	Consumer Protection Act
CTI	Comparative tracking index
DC	Direct current
EC	European Community
ECN	Engineering Change Note
EEA	European Economic Area
EEC	European Economic Community
EESR	Electrical Equipment (Safety) Regulations
EMC	Electromagnetic compatibility
EN	European Norme
EU	European Union
EUT	Equipment Under Test
FS	Final safety
HD	Harmonized document
HSE	Health and Safety Executive
IEC	International Electrotechnical Commission
ISO	International Standards Organization

ITE	Information technology equipment
LACOTS	Local Authority Co-ordination of Trading Standards
LVD	Low Voltage Directive
NCB	National Certification Body
OJ	*Official Journal*
PCR	Production Change Request
PROSAFE	Product Safety Enforcement Forum of Europe
PWB	Printed Wiring Board
QC/QA	Quality Control/Quality Assurance
SC	Safety Critical
SELV	Safety Extra Low Voltage
TCF	Technical Construction File
TF	Technical File
TSO	Trading Standards Officer
UKAS	United Kingdom Accreditation Service
UL	Underwriters Laboratory
UN	United Nations
V	Volts

Appendix 20

Templates

Templates and a report format that could assist the reader in creating his own self-assessment report and to set up a suitable production control system can be found in Appendices 20a to 20e (pp. 208–225), these are also available on a separately supplied 3.5″ floppy disk (for purchase details, see page 226).

Appendix 20a – Safety Critical Assembly List

This template can be used to list Safety Critical assembly operations; and it provides guidance for quality control or industrial engineers in preparing operation standards. It is particularly useful in cases where more than one product is manufactured on the site – refer also to Chapter 14.

Appendix 20b – Internal Audit Checklist

This is a simple template which can be used when performing audits in the plant, or for operations that need regular inspection – refer also to Chapter 14.

Appendix 20c – Product Evaluation Report

Chapter 11 explains in detail how to create and complete an evaluation report following the assessment of a product. Appendix 20c shows a blank template which can easily be modified to suit individual needs in the compilation of such reports. The report for each product category will be different – the report template offers a summary confirmation of the fundamental safety requirements but is by no means exhaustive.

Appendix 20d – Operation Worksheet

This template offers a typical operation worksheet format that may be used to provide instructions to the operator on how to perform a task – refer also to Chapter 14.

Appendix 20e – Safety Critical Components List

Safety Critical components need to be identified and listed so that all relevant areas of the operation can easily recognize them. Appendix 20e shows a template which may be used to list all such components – refer also to Chapter 14.

Other templates

Declaration of Conformity – Two templates of a DoC can be found in Appendix 13 – refer also to Chapter 13.

ECN – Engineering Changes Notes are very important in any responsible organization, they are used to control changes to material and components and to ensure that only approved changes are implemented in production, this template is a typical such document and can be found in Appendix 15 – refer also to Chapter 14.

PCR – Production Change Request is a similar template to the ECN but it is used mostly to instruct Production to apply changes in manufacturing process rather than changes to materials or components – refer also to Chapter 14. This template can be found in Appendix 16.

Appendix 20a

Safety Critical Assembly List

Task	Model	Comment
Issued by:	Issue no.:	Issue date:

Appendix 20b

Internal Audit Checklist

Audit date:	Auditor:	Signed:
AREA	**COMMENTS**	**Pass/Fail**

Appendix 20c

Production Evaluation Report

Report No: **Page 1 of**

PRODUCT EVALUATION REPORT	
Based on Safety Standard:	
Compiled by (+ signature):	
Checked by (+ signature):	
Date of issue:	
Evaluated by: Address:	
Type of equipment:	
Model/type reference:	
Manufacturer's name & address:	
Operating condition:	
Equipment mobility:	
Size (H × W × D):	
Mass (kg):	

Report No: **Page 2 of**

Phase (single/three):	
Protection against ingress of water:	
Classification:	
Current rating:	
Voltage rating (range):	
Power rating:	
Environment of use:	
Connection to the supply mains:	
Copy of marking plate and other labels:	
CONCLUSIONS:	

Report No: **Page 3 of**

Section 1

Section	Requirement – Test	Remarks	Results
		1.1.0 to 1.1.11 refer to the rating label	
1.1.0	MARKING and INSTRUCTIONS		
1.1.1	Legibility		
1.1.2	Durability		
1.1.3	Rated voltage (V)		
1.1.4	Rated power (W or VA)		
1.1.5	Rated current (A)		
1.1.6	Symbol for nature of supply		
1.1.7	Rated frequency (Hz)		
1.1.8	Symbol for Class II (if applicable)		
1.1.9	Equipment manufacturer		
1.1.10	Type/model		
1.1.11	Safety information		
1.1.12	Fuse holder		
1.1.13	Mains switch		
1.1.14	Earth symbol		
1.1.15	Marking for voltage/frequency settings		
1.1.17	Placing of markings		
1.1.18	Instructions when protection relies on building installation		
1.1.19	Letter or symbols used (according to IEC 417)		
1.1.20	Instructions for use – language		
1.1.21	Exclamation mark in triangle ⚠		
1.1.22			
1.1.23			

Report No: **Page 4 of**

Section 2

			Results
2.1.0	PROTECTION AGAINST HEATING HAZARDS – under NORMAL operating conditions	Remarks See Table 1	
2.1.2	Temperature rise on accessible parts		
2.1.3	Temperature rise on parts providing electrical isolation		
2.1.4	Temperature rise on parts acting as mechanical barriers		
2.1.5			
2.1.6			
2.1.7			
2.1.8			

Section 3

			Results
3.1.0	PROTECTION AGAINST HEATING and OTHER HAZARDS – under FAULT operating conditions	Remarks See Table 2	
3.1.1	Shock hazard		
3.1.2	Hazard from softening solder		
3.1.3	Measurement of temperature rise		
3.1.4	Temperature rise on accessible parts		
3.1.5	Temperature rise on PWBs		
3.1.6	Blocked fan rotors, vents, paper jam, etc.		
3.1.7			
3.1.8			
3.1.9			

Report No: Page 5 of

Section 4.1

4.1.0	PROTECTION AGAINST ELECTRIC SHOCK	Remarks See Tables 3, 4 & 5	Results
4.1.1	Protection against operator contact		
4.1.2	Internal wires		
4.1.3	Service area		
4.1.4	Shafts, knobs, handles, etc.		
4.1.5	Pins and plugs		
4.1.6	Access via ventilation holes		
4.1.7	Terminal devices		
4.1.8	Pre-set controls		
4.1.9	Removal of protective covers by hand		
4.1.10	Measurement of peak current		
4.1.11	Method of insulation		
4.1.12	Insulating materials		
4.1.13	Dielectric strength test		
4.1.14	Insulation resistance test		
4.1.15	Bridging of double or reinforced insulation		
4.1.16	Detaching of wires		
4.1.17			
4.1.18			
4.1.19			
4.1.20			

Report No: **Page 6 of**

Section 4.2

		Remarks	Results
4.2.0	PRIMARY POWER ISOLATION (requirement by some standards)		
4.2.1	Disconnection device		
4.2.2	Isolator used		
4.2.3	Disconnection of both poles simultaneously/all poles, etc.		
4.2.4	Marking of switch/isolator		
4.2.5			
4.2.6			
4.2.7			

Section 4.3

		Remarks	Results
4.3.0	OVERCURRENT AND EARTH FAULT PROTECTION IN PRIMARY CIRCUITS		
4.3.1	Number and location of protective devices		
4.3.2	Type of protective device and breaking capacity		
4.3.3			
4.3.4			
4.3.5			

Section 4.4

		Remarks	Results
4.4.0	PROTECTIVE EARTH		
4.4.1	Reliable connection		
4.4.2	Green/Yellow insulation		
4.4.3	Risk of corrosion		
4.4.4	Earth connector resistance		
4.4.5			
4.4.6			

Report No: **Page 7 of**

Section 4.5

4.5.0	EARTH LEAKAGE CURRENT (required by some standards)	Remarks See Table 6	Results
4.5.1	Test voltage applied		
4.5.2	Measured current (mA)		
4.5.3			
4.5.4			

Section 4.6

4.6.0	WIRING, CONNECTION TO SUPPLY	Remarks	Results
4.6.1	Type of connection		
4.6.2	AC inlet		
4.6.3	Construction and fixing of mains terminal		
4.6.4	Power cord and cross-sectional area		
4.6.5	Cord securing		
4.6.6	Risk of cord damage by bushing		
4.6.7			
4.6.8			

Section 5.1

5.1.0	PROTECTION AGAINST MECHANICAL HAZARDS	Remarks	Results
5.1.1	Vibration/Bump/Drop test		
5.1.2	Impact test		
5.1.3	Fixing of knobs, handles, etc.		
5.1.4	Stability test (required by some standards only)		
5.1.5	Edges and corners		

Report No: **Page 8 of**

Section 5.2

5.2.0	ELECTRICAL CONNECTIONS AND MECHANICAL FIXINGS	Remarks	Results
5.2.1	Cover fixings		
5.2.2	Fixing of the enclosure		
5.2.3	Fixing of internal wiring		
5.2.5	Fixing of uninsulated conductors		
5.2.6			
5.2.7			

Section 6

6.1.0	COMPONENTS	Remarks See Table 7	Results
6.1.1	Use of approved Safety Critical components		
6.1.2	Tests on non-approved Safety Critical components		
6.1.3			
6.1.4			

Report No: **Page 9 of**

Section 7

7.1.0	**PROTECTION AGAINST FIRE HAZARDS**	Remarks See Table 8	Results
7.1.1	Flammability of: • Materials • Components • Wiring harness • Enclosure • Decorative parts • PWB • etc.		
7.1.2	Protection against spread of flammable liquids		
7.1.3	Flammability of high voltage components		
7.1.4			
7.1.5			

Report No: Page 10 of

Section 8

8.1.0	PROTECTION AGAINST SPLASHING WATER	Remarks	Results
8.1.1	Use of appropriate enclosure		
8.1.2	Splash treatment		
8.1.3	Humidity treatment		
8.1.4			
8.1.5			

Section 9

9.1.0	PROTECTION FROM OTHER HAZARDS	Remarks	Results
9.1.1	• Implosion • Chemical • Energy • Radiation: 　– X-ray 　– laser 　– ultraviolet 　– microwave • Acoustic • Overflow • Spillage • Liberation of poisonous gases • Explosion • etc.		

Report No: **Page 11 of**

Section 10 – Tables

Table 1 Temperature rise (dT) measurement (normal condition)

Operating conditions:

Ambient temperature:

Part/location of temperature rise:	Test voltage	dT measured (°C)	Test voltage	dT measured (°C)	dT limit (C)

Table 2 Temperature rise (dT) measurement (fault condition)

Ambient temperature:

Component	Fault	Test voltage	Monitor point	Results (dT)

Report No: **Page 12 of**

Table 3 Dielectric strength test

Test voltage applied between:	Test voltage and test duration	Result (i.e. breakdown of insulation)

Table 4 Insulation resistance

Resistance measured between:	Applied voltage and test duration	Result (Ω)	Limit (Ω)

Table 5 Clearances (cl) and creepage (cr) distances

Pollution degree:

Distance measured between:	cl measured	cl required	cr measured	cr required	Result (pass/fail)

Report No: **Page 13 of**

Table 6 Earth leakage current

Operating condition	Test voltage	Measured current	Limit and result (pass/fail)

Table 7 List of Safety Critical components

Component	Reference number	Type or model	Manu-facturer	Rating	Standard	Approval

Table 8 List of flame retardant materials

Material or component, etc.	Flammability rating	Acceptable

Report No: **Page 14 of**

Section 11 – Photographs

Appendix 20d

Operation Worksheet

OPERATION WORKSHEET

| Title: | Issued by: | Model: | Issue no.: |
| | | | Date: |

Caution: This is a Safety Critical Operation

Appendix 20e

Safety Critical Components List

Part description	Value/ Rating	Identification	Model/ Location	Part no.	Supplier	Approval mark

Issued by:	Issue date:	Issue:

Template purchase details

The templates as shown in Appendices 20a–20e and Appendices 13, 15 and 16 can easily be customized to meet individual needs depending on the product manufactured. A 3.5″ floppy disk containing files with these templates can be purchased from *Gainspeed Ltd* as shown below. The files have been prepared in Microsoft Word (for Windows '95 or Windows NT).

Please return the order form to:

Gainspeed Ltd
33 Springfield Gardens
Bridgend, Mid Glamorgan
CF31 1NP
United Kingdom
Tel: +44 1656 650937
E-mail: jimtz@gainspeed.freeserve.co.uk
For credit cards fax on +44 1732 746617

Order form

Please send me _____ copies of the 3.5″ floppy disk titled 'Electrical Product Safety – A Step-by-Step Guide to LVD Self-Assessment – Templates' at £29 + P&P (£1.99 UK, £4.99 Europe, £5.99 Rest of World)

Name (include title and initials) _____

Date _____

Organization _____ Position _____

Address _____

_____ Post code _____

Tel. no. _____ Fax no. _____

E-mail _____

I enclose a cheque*/please debit my Visa/Mastercard for (total price):

£ _____

Card no. ☐☐☐☐ ☐☐☐☐ ☐☐☐☐ ☐☐☐☐

Expiry date: ☐☐ ☐☐   Signature _____

* Cheques should be made in £ (sterling) and payable to *Gainspeed Ltd*. A receipt will be dispatched with the disk. Please allow up to 28 days for delivery.

Index

Approval marks, 52
Audits, 132, 206

Basic insulation, 57, 61
Bridging insulation, 63
British Standards Institution, 26
Bump test, 74, 169

CE Marking, 3
Certification marks, 31, 51
CE Marking Directive, 4
Class I, 57, 58
Class II, 57, 60
Class III, 57, 60
Components, 39, 49, 50, 77, 170
Creepage, 57, 64, 181
Clearances, 57, 64, 181

Dangerous chemicals, 48
Dielectric Strength test, 81, 130, 168
Double insulation, 57, 62
Due Diligence, 15, 142

Earth Continuity Test, 130
Earth leakage current, 83, 131
EC Declaration of Conformity, 8, 118, 193
Electric shock, 45
Energy hazards, 47
Enclosures, 53, 76
Enforcement bodies, 12
Enforcement tools, 12
Engineering Change Note, 127, 196, 202

Equipment/Test instruments, 83, 161, 187
Evaluation Report, 40, 89, 206
European standards, 26
EU Directives, 146
    Fire, protection against, 46, 75, 174
    Fuses, 53, 77, 86, 171

Harmonized standards, 26, 163
Heating, 78, 79

Impact test, 75, 169
Insulation, 61, 175, 179
Insulation Resistance test, 82, 131
Interlocks, 54

Labels, 54, 85
Low Voltage Directive, 6

Mains cord/plug, 54, 68, 72, 170, 172
Market surveillance, 11
Markings, 85, 87, 164, 191
Mechanical hazards, 46, 74

National Certification Body, 33, 89, 156, 158

Operation standards, 128, 206

Policing Directives, 11
Pollution degree, 66

230   Index

Product specific standards, 25
Production control, 123
Production Change Request, 127, 199
Protective earthing, 58, 172

Radiation, 48, 53, 85, 164
Records, 132
Reinforced insulation, 57, 62

Safety Critical components/
    assemblies, 39, 49, 53, 124, 139,
    195, 206, 207
SMT, 33
Standards, 25, 37, 73, 148
Supplementary insulation, 57, 61

Technical File, 8, 115
Temperature/temperature rises, 47,
    55, 78, 165
TBM, 33
TMP, 33

User instructions, 55, 88

Vibration test, 74, 169

Wiring, 67